Composites Science and Technology

Series Editor

Mohammad Jawaid, Lab of Biocomposite Technology, Universiti Putra Malaysia,
INTROP, Serdang, Malaysia

Composites Science and Technology (CST) book series publishes the latest developments in the field of composite science and technology. It aims to publish cutting edge research monographs (both edited and authored volumes) comprehensively covering topics shown below:

- Composites from agricultural biomass/natural fibres include conventional composites-Plywood/MDF/Fiberboard
- Fabrication of Composites/conventional composites from biomass and natural fibers
- Utilization of biomass in polymer composites
- Wood, and Wood based materials
- Chemistry and biology of Composites and Biocomposites
- Modelling of damage of Composites and Biocomposites
- Failure Analysis of Composites and Biocomposites
- Structural Health Monitoring of Composites and Biocomposites
- Durability of Composites and Biocomposites
- Biodegradability of Composites and Biocomposites
- Thermal properties of Composites and Biocomposites
- Flammability of Composites and Biocomposites
- Tribology of Composites and Biocomposites
- Bionanocomposites and Nanocomposites
- Applications of Composites, and Biocomposites

To submit a proposal for a research monograph or have further inquries, please contact springer editor, Ramesh Premnath (ramesh.premnath@springer.com).

More information about this series at http://www.springer.com/series/16333

Mohamed Thariq Hameed Sultan ·
Ain Umaira Md Shah · Naheed Saba
Editors

Impact Studies of Composite Materials

 Springer

Editors
Mohamed Thariq Hameed Sultan
Department of Aerospace Engineering
Faculty of Engineering
Universiti Putra Malaysia
Seri Kembangan, Selangor, Malaysia

Ain Umaira Md Shah
Department of Aerospace Engineering
Faculty of Engineering
Universiti Putra Malaysia
Seri Kembangan, Selangor, Malaysia

Naheed Saba
Laboratory of Biocomposite Technology
Institute of Tropical Forestry and
Forest Products
Universiti Putra Malaysia
Seri Kembangan, Selangor, Malaysia

ISSN 2662-1819   ISSN 2662-1827 (electronic)
Composites Science and Technology
ISBN 978-981-16-1325-8   ISBN 978-981-16-1323-4 (eBook)
https://doi.org/10.1007/978-981-16-1323-4

This Springer imprint is published by the registered company Springer Nature Singapore Pte Ltd.
The registered company address is: 152 Beach Road, #21-01/04 Gateway East, Singapore 189721,
Singapore

Dedicated to my beloved family.

Preface

The present book is aimed to discuss the low and high-velocity impact properties of composites, which is known to be very important, mainly to the structural integrity and durability of materials. Impact, crash or collision can be the major reason lead to catastrophic failure as the load applied on the structure comes in a forcible way within short second. Aircraft structure is among the most susceptible to high-velocity impact events, due to the high relative velocities between an aircraft and the second objects. Bird strikes onto the radome of an aircraft is the easiest example of high-velocity impact events. Meanwhile, debris hitting the car bumper on the road is the example of low-velocity impact event.

Either high- or low-velocity impact events, the damages occurred in the structure needs in-depth and detailed observations to suggest the durability and service life of respective materials. Compared to metals, damages occurred in composites are more complex due to the combination of at least two different constituents to build its structure. There are many challenges in detecting damages in composites compared to metals. The presence of different structures, which physically separated to one another, leads to a difficult analysis of damages in composites.

There are two main techniques in detecting and analyzing damages in composites, which are non-destructive and destructive techniques. Non-destructive techniques are the most commonly used in the industries as it does not affect the overall structure of materials. The examples of non-destructive techniques are ultrasonic, radiography, magnetic particle, acoustic emission and dye penetrant. Some of these techniques can provide a complete volumetric inspection of a structure, while some techniques only provide surface inspection, depending on the types of materials.

The analysis of low- and high-velocity impact properties of composites can also be carried out through numerical analysis. Prediction on the condition of composites after an impact event can be suggested based on the numerical analysis using finite element analysis. Very less study was reported on this scope of numerical analysis on composites, since the standard properties of composites are inconsistent due to its custom composition during each fabrication. However, researchers have started to explore this area to provide more choices to the industries in predicting the strength and service life of their materials, without having the experimental analysis.

In overall, we hope that the discussion presented in this book will provide better understanding to the reader on the impact properties of composites, through both the experimental and numerical analyses. Damage analysis is very important in composites to avoid sudden failure, either due to high-velocity impact or multiple low-velocity impact events. We are very grateful to have all the authors to share their findings and expertise in this book. Lastly, we would also thank the Springer team for the support and guidance in completing this book.

Seri Kembangan, Malaysia

Ain Umaira Md Shah
Mohamed Thariq Hameed Sultan
Naheed Saba

Contents

Contributors

Nur Marini Zainal Abidin Department of Aerospace Engineering, Faculty of Engineering, Universiti Putra Malaysia, Serdang, Selangor, Malaysia

Adi Azriff Basri Department of Aerospace Engineering, Faculty of Engineering, Universiti Putra Malaysia, Serdang, Selangor, Malaysia

Ernnie Illyani Basri Department of Aerospace Engineering, Faculty of Engineering, Universiti Putra Malaysia, Serdang, Selangor, Malaysia

Yi Di Boon School of Mechanical and Aerospace Engineering, Nanyang Technological University, Jurong West, Singapore

M. T. H. Sultan Aerospace Manufacturing Research Centre (AMRC), Faculty of Engineering, Universiti Putra Malaysia (UPM), Serdang, Selangor Darul Ehsan, Malaysia;
Department of Aerospace Engineering, Faculty of Engineering, Universiti Putra Malaysia, Serdang, Selangor Darul Ehsan, Malaysia;
Laboratory of Biocomposite Technology, Institute of Tropical Forestry and Forest Products (INTROP), Universiti Putra Malaysia, Serdang, Selangor Darul Ehsan, Malaysia;
Aerospace Malaysia Innovation Centre (944751-A), Prime Minister's Department, MIGHT Partnership Hub, Cyberjaya, Selangor Darul Ehsan, Malaysia

M. Jawaid Laboratory of Biocomposite Technology, Institute of Tropical Forestry and Forest Products (INTROP), Universiti Putra Malaysia, Serdang, Selangor Darul Ehsan, Malaysia

Sunil Chandrakant Joshi School of Mechanical and Aerospace Engineering, Nanyang Technological University, Jurong West, Singapore

Chandrakant R. Kini Department of Aeronautical and Automobile Engineering, Manipal Institute of Technology, Manipal Academy of Higher Education, Manipal, Karnataka, India

S. H. Lee Laboratory of Biopolymer and Derivatives, Institute of Tropical Forestry and Forest Products (INTROP), Universiti Putra Malaysia, Serdang, Selangor, Malaysia

A. A. Mazlan Department of Aerospace Engineering, Faculty of Engineering, Universiti Putra Malaysia, Serdang, Selangor, Malaysia

Suhas Yeshwant Nayak Department of Mechanical and Manufacturing Engineering, Manipal Institute of Technology, Manipal Academy of Higher Education, Manipal, Karnataka, India

Noorshazlin Razali Department of Aerospace Engineering, Faculty of Engineering, Universiti Putra Malaysia, Serdang, Selangor, Malaysia

N. Saba Laboratory of Biocomposite Technology, Institute of Tropical Forestry and Forest Products, Universiti Putra Malaysia, Serdang, Selangor, Malaysia

S. N. A. Safri Laboratory of Biocomposite Technology, Institute of Tropical Forestry and Forest Products, Universiti Putra Malaysia, Serdang, Selangor, Malaysia

Rashmi Samant Department of Mechanical and Manufacturing Engineering, Manipal Institute of Technology, Manipal Academy of Higher Education, Manipal, Karnataka, India

S. Sani Industrial Technology Division, Malaysian Nuclear Agency, Kajang, Selangor, Malaysia

I. Sankar Department of Mechanical Engineering, National Engineering, Kovilpatti, Tamil Nadu, India

Fabrizio Sarasini Department of Chemical Engineering Materials Environment, Sapienza-Università Di Roma, Roma, Italy

Francesca Sbardella Department of Chemical Engineering Materials Environment, Sapienza-Università Di Roma, Roma, Italy

Chithirai Pon Selvan School of Science and Engineering, Curtin University Dubai, Dubai, United Arab Emirates

Claudia Sergi Department of Chemical Engineering Materials Environment, Sapienza-Università Di Roma, Roma, Italy

A. U. M. Shah Department of Aerospace Engineering, Faculty of Engineering, Universiti Putra Malaysia, Serdang, Selangor Darul Ehsan, Malaysia;
Laboratory of Biocomposite Technology, Institute of Tropical Forestry and Forest Products (INTROP), Universiti Putra Malaysia, Serdang, Selangor Darul Ehsan, Malaysia

B. Satish Shenoy Department of Aeronautical and Automobile Engineering, Manipal Institute of Technology, Manipal Academy of Higher Education, Manipal, Karnataka, India

K. Rajath Shenoy Department of Mechanical and Manufacturing Engineering, Manipal Institute of Technology, Manipal Academy of Higher Education, Manipal, Karnataka, India

Avinash Shinde Department of Mechanical Engineering, Cummins College of Engineering, Pune, Maharashtra, India

I. Siva Department of Mechanical Engineering, Centre for Composite Materials, Kalasalingam Academy of Research and Education, Anand Nagar, Tamil Nadu, India

S. I. B. Syed Abdullah Department of Aeronautics, Imperial College London, London, UK

K. Tabrej Department of Aerospace Engineering, Faculty of Engineering, Universiti Putra Malaysia, Serdang, Selangor Darul Ehsan, Malaysia

Jacopo Tirillò Department of Chemical Engineering Materials Environment, Sapienza-Università Di Roma, Roma, Italy

Low Velocity Impact Testing on Laminated Composites

S. I. B. Syed Abdullah

Abstract The response of composite laminates from transverse impact loading is known to vary with the speed of impact. In Low Velocity Impact (LVI) conditions, boundary effects usually dominate since the impact duration is longer between the laminate and the impactor. The global damage modes in LVI is also distinctly unique, whereby large deflections often occur, which depend highly on the shear properties (both in-plane and interlaminar) of the material. Therefore, characterisation of impact resistance and damage on LVI conditions are crucial before material selection for structural design. In this chapter, the LVI behaviour of composite laminates under LVI loading is investigated. The type of damage under LVI is also highlighted and discussed to obtain a detailed understanding of the impactor mass and velocity effects. The extent of delamination is studied using ultrasonic C-scan and radiograph images. Finally, where possible, fractographic studies have been undertaken to understand the influence of the interlaminar toughness on the impact resistance.

Keywords Low velocity impact · Impact resistance · Fracture toughness · Delamination · Impact damage

1 Impact on Composite Structures

The problem of impact on composite structures has been a subject of review for more than three decades. To date, many review papers have been written by researchers (Davies and Olsson 2004; Richardson and Wisheart 1996; Argawal et al. 2014; Abrate 1991a, b, 1998; Vaidya 2011; Silberschmidt 2016; Cantwell and Morton 1991) reporting the advances observed in the field of impact mechanics on composite materials. These advances, which include those made in damage prediction using numerical methods such as FEM, have strengthened our understanding of more damage tolerant structures, designed for various applications. Mathematical models such as the spring-mass model, the energy-balance model and the Delamination Threshold Load (DTL) (Schoeppner and Abrate, 2000a, b; Donadon and Falzon 2006) have

S. I. B. Syed Abdullah (✉)
Department of Aeronautics, Imperial College London, London, UK

© Springer Nature Singapore Pte Ltd. 2021   1
M. T. H. Sultan et al. (eds.), *Impact Studies of Composite Materials*, Composites Science and Technology, https://doi.org/10.1007/978-981-16-1323-4_1

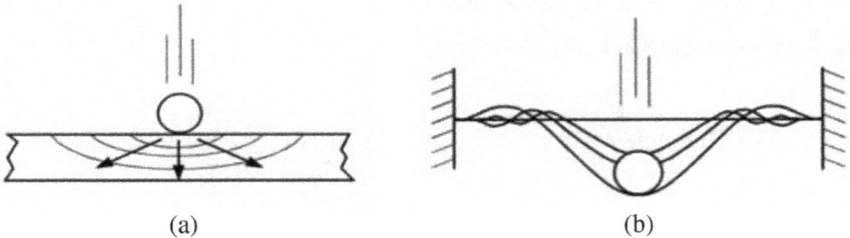

Fig. 1 Response types for impacted composite structures, **a** Stress wave dominated event, **b** Boundary dominated event (Davies and Olsson 2004)

greatly helped in determining the performance of the composite structure under transverse impact loading. Foreign Object Damage (FOD), characterised by the velocity of impact, is of importance due to its relevance in real-life applications. Research normally centres around two types of velocity regime—low and high velocity impact. Although argument exists with regards to the distinction of velocity range between low and high velocity impact, researchers have agreed that the threshold velocity defining low velocity impact is up to 10 m/s^{-1} (Davies and Olsson 2004; Richardson and Wisheart 1996; Cantwell and Morton 1991; Sjoblom and Hartness 1988), and more generally by Abrate (1998) for velocities below 100 m/s^{-1}. Additionally, Davies and Robinson (Robinson and Davies 1992) define Low Velocity Impact (LVI) as one in which through-thickness stress waves in the specimen play no significant part in the stress distribution at any time during the impact event. Hence, a global deformation can be observed in the laminate, shown in Fig. 1b, due to the long impact duration. In contrast, a High Velocity Impact (HVI) event is usually stress wave dominated, Fig. 1a, therefore the effects of boundary conditions can be neglected (Davies and Olsson 2004; Abrate 1991a, b, 1998).

Godwin and Davies (1988) proposed a simple technique to evaluate the transition velocity in which stress wave effects dominate. The relationship is based on the propagation of stress waves from the front face, which then progress towards the rear of the laminate. Therefore, the compressive strain, ε_c, can be calculated by using the relationship given by Robinson and Davies (1992):

$$\varepsilon_c = \frac{V_i}{C_z} = \frac{V_i}{\sqrt{\frac{E_z}{\rho}}} \tag{1}$$

where V_i is the impact velocity, C_z is the through-thickness speed of sound in the material, V_i is the out-of-plane Young's modulus, and ρ is the material's density. For instance, the transition velocity into a stress wave dominated event would occur at an impact velocity of approximately 20 ms^{-1} and above for a Uni-Directional (UD) CF/Epoxy composite with $E_y = E_z = 9.4$ GPa, $\varepsilon_c = 0.6\%$, and $\rho = 1.3$ g/cm^3.

Under High Velocity Impact (HVI) load, the longitudinal and transverse stress waves propagate from the point of impact upon initial contact between the projectile and the composite. The longitudinal wave propagates towards the edge of the laminate and is reflected back towards the centre, whilst the transverse wave propagates towards the composite back face, usually travelling at a velocity much lower than the longitudinal wave. The repeated reflections of the longitudinal stress waves typically decrease in intensity throughout the impact event, which was shown experimentally by Pandya et al. (2008).

When the impact velocity (or impact energy) is considerably lower than the V50 (or penetration energy for LVI loading), minimal to no damage will be observed in the laminate. V50 is defined as the velocity at which a projectile has a 50% probability of penetrating the target material (in this case composite laminates) (Cuniff 1999a, b). The contact stress between the projectile (or impactor) and the laminate would generally induce minimal damage in the form of matrix cracking (Cantwell and Morton 1989a, b). These cracks would saturate as a result of coalescence between multiple microcracks (Olsson 2001; Berthelot 2003; Puck and Schurman 1998; Williams et al. 2003). Depending on the thickness of the laminate, the damage pattern would differ due to the difference in the energy absorbing mechanism. For a typical brittle fibre-matrix system such as Carbon Fibre/Epoxy (CF/Epoxy) and Glass Fibre/Epoxy (GF/Epoxy), the cross-sectional damage normally resembles the so-called "pine tree" pattern. For thin laminates, a reversed pine tree pattern, Fig. 2a, can usually be observed from the cross-section of the impact area. This is due to the high bending load at the rear side of the laminate, hence initiating shear matrix failure. On the contrary, matrix cracks in thick laminates will often result in a pine tree pattern, Fig. 2b. This is because the normal in-plane stresses have exceeded the transverse tensile stress of the front plies, therefore initiating matrix failure. For both pine tree patterns, extensive experimental evidence exists for CF/Epoxy composites (Cantwell and Morton 1985; Jih 1993; de Freitas et al. 2000; Bouvet et al. 2009; Garcia-Rodriguez et al. 2000), and GF/Epoxy (Zhou 2003; Shyr and Pan 2003; Liaw and Delale 2007; Crupi et al. 2016).

The saturation of matrix cracks would typically lead to interlaminar failure, often known as delamination damage, which severely degrades the laminate's capability to carry further loads. An obvious drop in the load-time history (or load–displacement history) response (Schoeppner and Abrate, 2000a, b) can be observed indicating a delamination has occurred. For some materials (usually brittle materials such as CF/Epoxy and GF/Epoxy), violent oscillations may also be observed following

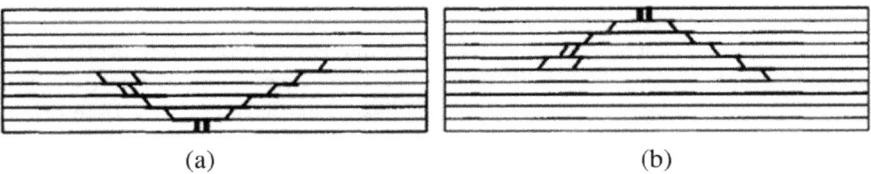

(a)   (b)

Fig. 2 **a** Reversed pine tree pattern, **b** Pine tree pattern (Abrate 1998)

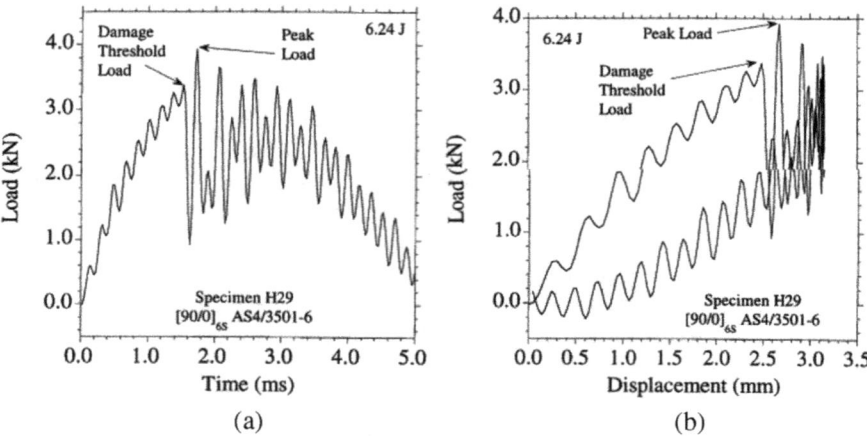

Fig. 3 Delamination Threshold Load (DTL) for a typical carbon/epoxy laminate, **a** Load-time history, **b** load–deflection history. Schoeppner and Abrate (2000a, b)

delamination damage, Fig. 3, due to the unstable propagation of delamination [24]. Also, a noticeable change in stiffness can be seen in the load–displacement history [20], illustrated clearly in Fig. 3b.

A closed form solution for the prediction of delamination damage has also been defined by Davies and Robinson (1992), utilising fracture mechanics given by:

$$P_c^2 = \frac{8\pi^2 E_f h^3}{9(1 - v^2)} G_{IIc} \tag{2}$$

where P_c is the critical load to initiate delamination, G_{IIc} is the Mode II critical strain energy, v is the Poisson's ratio, E_f is the laminate flexural modulus, and h is the laminate thickness. The high dependence on the shear properties for LVI loading is apparent, resulting in the inclusion of Mode II critical strain energy in Eq. (2). This is not surprising since LVI is a flexural dominated event, hence interlaminar shear properties are central to the impact performance of laminated composites. A closed form approximation has also been derived by Olsson et al. (2006; Olsson 2010) to predict delamination initiation and propagation under HVI loading. It was found that the threshold load to initiate delamination is approximately 21% higher compared to its quasi-static value. This is due to the dynamic nature of the HVI event, inducing very high contact stresses localizing at the point of impact (Olsson et al. 2006).

It must be noted that delamination failure normally occurs only at interfaces with different fibre orientation, Fig. 4a (Richardson and Wisheart 1996; Abrate 1998; Liu 1988). This is associated with the adjacent plies which have different fibre orientations possessing different bending stiffnesses, therefore promoting delamination failure due to the property mismatch, Fig. 4b. From this, peanut-shaped delamination extending along the fibres can normally be observed, increasing in size with

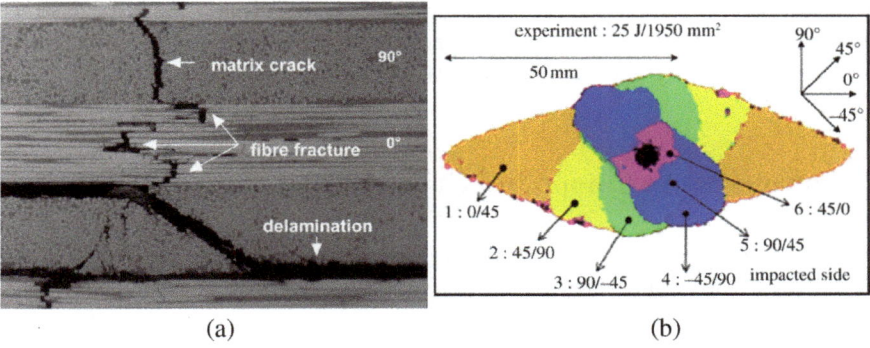

(a)   (b)

Fig. 4 **a** Variation of damage in low velocity impact loading (Davies and Olsson 2004), **b** Peanut-shaped delamination with respect to fibre orientation (Wisnom 2012)

an increase in the misalignment angle (i.e. a larger size should be present for a 0°/90° alignment compared to a 0°/45° fibre alignment) (Davies and Olsson 2004; Richardson and Wisheart 1996; Abrate 1998; Wisnom 2012).

If the velocity is increased (i.e. closer to the laminate ballistic limit or penetration energy), further damage would be seen due to higher contact stresses as well as a greater amount of energy transfer between the projectile and the laminate. For thin laminates, the energy transfer will normally result in tensile straining of the fibres, usually known as a cone formation at the rear of the laminate (Naik and Shrirao 2004; Shaktivesh et al. 2013; Naik et al. 2005). This is due to the high flexural load experienced by the laminate, with high tensile strain, particularly at the laminate back face. Penetration of the laminate is therefore governed by the amount of fibre failure; full penetration indicates that all fibres are broken, and no penetration indicates that zero or a minimal number of fibres have failed (Naik and Shrirao 2004). Fibre failure normally initiates from the laminate back-face, extending through-the-thickness towards the front face (Davies and Olsson 2004; Richardson and Wisheart 1996; Abrate 1998). For thick laminates, penetration is normally associated with shear failure, specifically in the transverse direction resulting in a shear-plug formation (Richardson and Wisheart 1996). The depth of penetration depends on the severity of damage and includes other factors such as the type and shape of the impactor (Richardson and Wisheart 1996; Abrate 1998; Cantwell and Morton 1991; Shyr and Pan 2003; Mitrevski et al. 2015, 2006). The energy related to laminate penetration is proposed by Dorey (1987) given as:

$$E_p = \pi \gamma t d \tag{3}$$

where E_p is the penetration energy, γ is a coefficient property related to the fibre fracture energy, d is the diameter of the impactor, and t is the laminate thickness. It is clear from Eq. (3), that the energy absorption due to penetration increases with increasing thickness. This has been confirmed experimentally by researchers (Richardson and

Wisheart 1996; Abrate 1998; Shyr and Pan 2003; Cantwell and Morton 1989a, b; Caprino and Lopresto 2001), and have concluded that the energy absorption (hence impact resistance) significantly increases with increasing thickness.

2 Polymer Fibre Composites Under Impact

The use of a polymer fibre as an alternative to conventional brittle fibres such as CF and GF offer a potential solution to the poor impact performance of composite materials. Its superior mechanical properties often result in an enhanced impact performance, both under LVI and HVI. A key step in understanding the impact performance of high-performance fibre composites was proposed by Cuniff (1999a, b) using a dimensionless analysis based on the fibre specific toughness and longitudinal strain wave velocity, given by:

$$U^* = \frac{\sigma \varepsilon}{2\rho} \sqrt{\frac{E}{\rho}} \tag{4}$$

where U^* is the so-called Cuniff velocity", σ is the fibre ultimate axial tensile strength, ε is the fibre ultimate tensile strain, ρ is the fibre density, and E is the fibre tensile modulus. Note the linear assumption in Eq. (4), which could be misleading for fibre (or composite) possessing a non-linear tensile response. Moreover, through-thickness stresses, as well as the in-plane shear properties were not considered in Eq. (4).

This may present a significant difference in the composite impact performance if the selection is based purely on U^*. However, Eq. (4) does allow for a quick analysis in selecting the appropriate fibre for composite design. Table 2.1 presents the calculated Cuniff velocity for selected commercially available fibres.

From Table 1, it can be easily observed that IM7 possesses the highest U^* when compared to the selected fibres. This should indicate that IM7 should possess excellent impact performance although in reality, this is not the case. It was found that IM7 composite was the worst performing composite when compared to Vectran and S2-Glass composites. This result is associated with several factors. First, the low tensile strain-to-failure of IM7 ($\approx$1.9% (Hexcel 2018)) compared to Vectran ($\approx$3.8% (Kuraray America 2006)) and S2-Glass (5.7% (AGY 2006)) limits the 'elongation' of the fibre to prevent impactor penetration. A recent investigation by Heimbs et al.

Table 1 Cuniff velocity for chosen fibres (Syed Abdullah 2019, 2021)

Fibre type	U^* (m/s)
IM7 (CF)	722
S2-Glass (GF)	634
Vectran	616

(2018) found that Dyneema HB26 showed superior HVI performance compared to CF and GF composites, whilst a comparable LVI performance with GF composite. Furthermore, in both impact events (LVI and HVI), CF composite exhibits the worst impact performance among all three composite materials. The superior performance can be partly attributed to the relatively high tensile strain-to-failure of Dyneema HB26 laminate, in addition to other factors such as the low $\pm 45°$ in-plane shear properties (both stiffness and strength). This is consistent with the works of Reddy et al. (2017) when investigating the impact performance (LVI and HVI) of GF/Epoxy and Dyneema HB26 laminates. The influence of shear strength on the ballistic performance has also been investigated by Karthikeyan et al. (2013), which found a significantly higher V_{50} for the uncured IM7/8552 (CF/Epoxy) if compared to its cured counterpart. More importantly, the low $\pm 45°$ in-plane shear properties have resulted in a superior impact performance of Dyneema HB26 and HB50 laminates; at least 50% higher than the uncured IM7/8552.

Recently, Hazzard et al. (2017) investigated the effect of fibre orientation of Dyneema HB26 under LVI loading. A large Back Face Deflection (BFD) due to the low $\pm 45°$ in-plane shear properties were observed for the $0°/90°$ orientation of Dyneema HB26 laminates. However, no penetration was observed even for large impact energies of up to 150 J impacted on a 2 mm thick laminate. The effects of fibre orientation of Dyneema HB26 under HVI loading have also been investigated by Karthikeyan et al. (2016), whereby a large BFD was observed for the $0°/90°$ orientation compared to other types of layup such as Unidirectional (UD) and helicoidal (hybridised $0°/90°$ orientation). In spite of this, the $0°/90°$ orientation yielded the highest V_{50} compared to the other orientation types. It must be noted that a large BFD may not necessarily be a positive characteristic concerning composite structural design. In certain structural design, particularly in defence applications, a large BFD may imply a severe weakness due to the 'soft' nature of the structure, enabling rapid (and excessive) membrane loading at the point of impact.

Secondly, the inherently large fracture energy of polymer fibre composites partly contributes to its superior impact performance. An earlier work by Park and Jang (2004) highlights the considerably large impact absorption energy of Aramid when compared to GF composites (6.3 J and 58.43 J for a 2 mm thick GF and Aramid laminates, respectively) under LVI loading. A similar observation was seen by Sikarwar and co-workers (2017) when investigating the behaviour GF/Epoxy and Kevlar/Epoxy laminates, where a considerably higher specific energy absorption capacity was found for Kevlar/Epoxy laminates ($\approx 36\%$ higher) when compared to Glass/Epoxy laminates. This is consistent with the findings of Evci and Gulgec (2012), where a superior impact performance was observed for Aramid laminates if compared to GF composites under LVI.

The high fracture energy of polymer fibres (or polymer composite systems) enables a large energy absorption under impact loading, partly due to its unique fibre structure. For instance, the so-called 'skin–core' structure of Vectran fibre typically results in a tougher system, due to its inherent energy absorption mechanisms such as fibre pull-out and interfacial sliding. A recent investigation on the tensile fracture properties of Vectran/Epoxy laminates (Syed Abdullah 2018) revealed a much higher value of fracture toughness; up to 48.26% and 95.27% higher for initiation and propagation for some CF/Epoxy (Pinho and Robinson 2012) composite system, and 9.93 and 68.6% higher for initiation and propagation for some GF/Epoxy (Katafiasz et al. 2019) composite system. The high fracture energy of Vectran is partly attributed to the 'skin–core' nature of the fibre, resulting from its manufacturing process.

During fibre extrusion, the 'skin' cools much quicker than the 'core' resulting in a highly oriented crystalline region at the 'skin', whilst a less oriented, amorphous region at the 'core'. Hence, a noticeable difference in the mechanical properties may be found, whereby the 'skin' is much stiffer when compared to the 'core', resulting in a path of failure which initiated first at the 'skin' and then propagating towards the 'core'.

Thirdly, the sensitivity to strain-rate effects was evident, particularly in polymer fibre composites. A recent report by Singh (2018) attempts to collate existing strain-rate research on high performance fibre composites. It was found that GF (or GF composites) display considerable sensitivity towards strain-rate, with enhancements close to 200% observed at 103/s compared to the quasi-static case. A similar response was seen for Kevlar fibres, although its enhancement is not as significant as GF, with an increase of up to 50% in tensile strength. This is consistent with the works of Wang and Xia (1998), when investigating the effect of strain-rate (0.0001/s–1350/s) on Kevlar 49 fibre bundles. The authors found an approximately 22% increase on its tensile strength modulus and strength, including a 15% increase in its tensile strain-to-failure. Tan et al. (2010) investigated the tensile response of Twaron CT716 yarns at strain-rates of up to 480/s and found a 28 and 36% increase on the tensile modulus and strength, respectively. In a follow-up study, Koh et al. (2010) investigated the strain-rate effects of Spectra 900 yarns and found a 150% increase in the tensile modulus and 40% increase in the tensile strength.

Despite the excellent impact performance of polymer fibres, the significantly low compressive properties severely restrict their use as a structural component. This is due to the weak (van der Waals) bonds between molecules, which easily fails when the fibres are loaded in compression. Specifically for Vectran, the fibrillar nature of the fibre tends to de-fibrillate almost instantly under compression loads, resulting in a compressive property of approximately 10% of its tensile component (Donald et al. 2006). Hence, it is common that polymer fibre composites should be utilised as a reinforcement in a hybrid composite system.

Park and Jang (2001, 2004) investigated the impact performance of Aramid fibre/glass fibre hybrid composite and found a markedly higher impact performance when compared to the monolithic GF composite system. Sikarwar et al. (2017) observed a 20% increase in the HVI performance of GF/Kevlar hybrid composites when compared to the monolithic GF composite system. Bandaru et al. (2001) investigated the ballistic performance hybrid thermoplastic Kevlar and basalt composite system and found a 26.27% increase in the impact performance compared to Basalt monolithic composite system. A recent review by Iannucci (2018) highlights the advantages of composite hybridisation by incorporating polymer fibres with conventional materials such as CF and GF. Some of the benefits include potential weight savings and enhanced impact performance, whilst achieving the desired compressive properties to be used as a structural component.

3   Post Impact Residual Strength

The post-impact residual strength of composites, also known as the damage tolerance, is an important parameter when designing structures involving composite materials. This is important since the damaged composite is expected to maintain its original strength and stiffness. The damage tolerance of a composite is usually studied by determining the effect of different impact energies on their residual strength, the Compression After Impact (CAI) test being the experimental test of components damaged under LVI loads. First, the composite is subjected to an LVI test, using a drop weight impact tower which records the load and impactor vertical displacement. Next, the damaged composite is placed in a bespoke rig, usually following a specified dimension from the ASTM D7137 standard (American Standard for Testing Materials (ASTM) 2017). It must be noted that other tests exist to quantify the composite damage tolerance, such as the Tension After Impact (TAI) (El-Zein and Reifsnider 1990), although the test is rarely performed due to the difficulties associated with the test.

Sanchez-Saez et al. (2005) investigated the CAI behaviour of thin CF/Epoxy laminate (1.6–2.2 mm thick) with three different laminate layups (QI, CP, and Woven), and found that the woven laminate possesses the highest CAI strength compared to the other layups. Khondker et al. (2005) studied the CAI behaviour of weft-knit architecture and concluded that the weft-knit composite exhibits a considerably superior CAI performance, if compared to UD and braided architecture. This is mainly attributed to the enhanced structural integrity of the weft-knitted fabric, resulting in a composite with an improved impact resistance thus suffering minimal damage during initial impact. Hart et al. (2017) compared the damage tolerance of 2D and 3D woven Glass/Epoxy composite under CAI and Flexure After Impact (FAI) loading and found that the FAI testing yields 70% reduction in flexural strength compared to only 20% reduction in compressive strength. In addition, it was found that the 3D woven composite possesses a superior damage tolerance due to the presence of the through-thickness stitches (Z-binder) in the composite, significantly improving its delamination behaviour. Tan et al. (2012) studied the effect of through-thickness stitching in CF/Epoxy composite and concluded that superior CAI performance was observed for composites having through-thickness stitches, especially those with high density stitches. This is mainly because of the enhancement in the delamination behaviour, due to the presence of through-thickness stitches in the composite.

4   Factors Influencing the Impact Performance

5   The Fibre Constituent

Perhaps the most important factor in determining the performance of a composite is its fibre constituent. The choice of fibre as a component in a composite greatly determines its impact performance, since the fibre is the primary load bearer in a laminate. For instance, the use of fibres with a large tensile strain-to-failure would often result in a better impact performance. This is since the large tensile strain-to-failure enables a larger energy absorption by way of a large BFD, hence preventing impactor/projectile penetration (Hazzard et al. 2017). This is shown in Figs. 5 and 7, where it can be observed that Vectran/MTM57 exhibits a much larger BFD compared to S2-Glass/MTM57. In addition to a large tensile strain-to-failure, the polymeric nature of Vectran fibres results in a lower ±45° in-plane shear properties (stiffness and strength), thus promoting a large BFD to occur. Under LVI loading, the interlaminar fracture behaviour is important, particularly the Mode II (G_{IIc}) fracture toughness. This is because under LVI loading, flexural deformation dominates hence laminates with a low G_{IIc} would outperform those with a higher G_{IIc}.

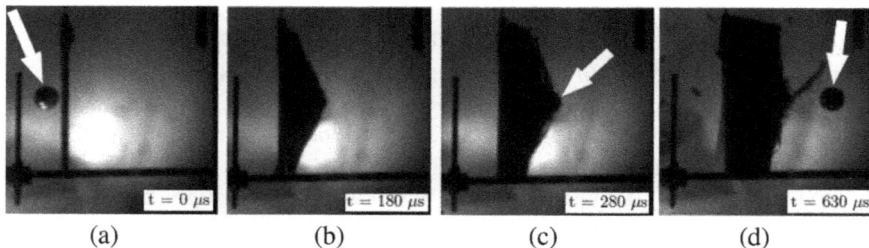

     (a)   (b)   (c)   (d)

Fig. 5 Montage of HVI events on Vectran/MTM57 at 171.2 m/s, **a** before impact, **b** during impact, **c** partial penetration, **d** penetrated. White arrow indicates projectile (Syed Abdullah 2019, 2021)

     (a)   (b)   (c)   (d)

Fig. 6 Montage of HVI events on S2-Glass/MTM57 at 183.08 m/s, **a** before impact, **b** during impact, **c** partial penetration, **d** penetrated. White arrow indicates projectile (Syed Abdullah 2019, 2021)

6 The Matrix Constituent

It has been shown that the use of thermoplastic matrix results in an improved impact performance if compared to its thermosetting matrix counterpart. This is because there is generally no crosslinking in thermoplastic matrices when heated above its liquid to glass temperature (T_g). In fact, compared to thermosetting matrices, thermoplastic matrix will soften when heated above T_g and harden when cooled down. This mechanism is usually repeatable, whereby thermoplastic materials can be reheated again and formed into desired shape. In contrast, thermosetting materials will harden when heated above its critical temperature and will not soften again on reheating (Cowie 1991). Therefore, thermoplastic materials are often tougher compared to thermosetting materials.

Vieille et al. (2013) compared the LVI performance of CF/Epoxy (thermosetting matrix) and CF/PEEK (thermoplastic matrix) and found that the CF/PEEK composite performed considerably better compared to its CF/Epoxy counterpart. Arikan and Sayman (2015) investigated the impact behaviour of E-glass fibre reinforced Polypropylene and epoxy matrix composites and found that the composite with thermoplastic composites performed considerably better under LVI loading. In addition, the 'ductile' nature of the thermoplastic matrix greatly reduces delamination damage from occurring. Dorey et al. (1985) compared the LVI performance of epoxy and Polyetherketone (PEEK) matrix with carbon fibre reinforcement. While damage were less extensive for CF/PEEK composites, it was found that the damage in the carbon fibre/PEEK composite comparable to that of carbon fibre/epoxy. Recently, Sonnenfeld et al. (2017) investigated the use of thermosetting-thermoplastic combination with carbon fibre reinforcement and found a significant reduction in damage due to LVI loading.

7 Fabric Architecture and Layup Orientation

The fabric architecture of a composite could greatly contribute to its impact performance. For instance, Non-Crimp Fabrics (NCF) will have a superior impact behaviour if compared to Uni-Directional (UD) composites, although its layup orientation is comparable. This is because the stitches present in NCF based composite will significantly improve its delamination behaviour, resulting in a superior LVI performance (Greenhalgh 2009). In addition, the improvement in its delamination behaviour is particularly beneficial due to the dominance of flexural deformation in LVI conditions.

The effect of layup orientation is also influential in determining the impact performance of a composite. Depending on the layup, the orientation may contribute to the flexural behaviour of the composite. Under impact loading, a more 'compliant' composite is desirable to absorb the impact energy transferred from the impactor/projectile to the laminate. This is usually done by incorporating a higher number of ±45° plies in the laminate. As a result, the laminate will be less stiffer in flexure thus able to deform further to prevent penetration from occurring. Hazzard et al. (2017) investigated the effect of three different layup orientation (cross-ply, quasi-isotropic, and helicoidal) and found that the helicoidal orientation has the most superior LVI performance compared to the other two layups. Note that Cross-Ply (CP) laminates are obtained when each plies are arranged in a 0° and 90° only, whilst the Quasi-Isotropic (QI) layup is defined as a laminated in which the plies are arranged in such a way that the in-plane properties will behave as an isotropic composite, though the through thickness properties is not isotropic. Sharma et al. (2019) found that damage exerted by a Cross-Ply (CP) layup is considerably higher compared to a Quasi-Isotropic (QI) layup under LVI loading. In addition, the QI layup was also able to absorb a higher impact energy under LVI loading. Zhou et al. (2019) studied the effect of different stacking sequence and ply orientation on CF/Epoxy composite and found a significant dependence in the stacking sequence and ply orientation on the LVI performance. It was found that the QI layup is considerably superior compared to the CP layup under LVI loading, consistent with the findings of the previous researchers.

As mentioned earlier, the LVI performance of composite laminates is highly dependent on its fabric architecture. Recently, Miao et al. (2019) studied the effect of three different fabric architecture (UD, Woven, and 3D composite) and found that the 3D composite possesses better impact resistance compared to the UD and Woven composite. This is because the Z-binder which is present in a 3D composite act as initiation site due to the weak fibre-matrix bonding, resulting in an enhanced impact resistance for the composite. Similarly, Saleh et al. (2019) investigated the impact performance of three different fabric architecture (NCF, 2D and 3D woven composite) under multiple LVI and found that damage was the least in 3D woven composite. In addition, the residual strength under CAI loading was the highest for 3D woven composite if compared to the other two laminates used (NCF and 2D woven). A similar observation was also found by Syed Abdullah (2019, 2021), when investigating the LVI performance of three different monolithic composite (IM7/8552, S2-Glass/Epoxy, and Vectran/Epoxy) having similar weight (≈90 g) using two different fabric architecture (NCF and UD). It was found that both S2-Glass/Epoxy and Vectran/Epoxy possesses superior impact properties compared to the IM7/8552. This was due to several reasons. First, both S2-Glass/Epoxy and Vectran/Epoxy composite has a relatively larger tensile strain-to-failure compared to IM7/8552 (2.6% and 2.8% respectively for Vectran/Epoxy and S2-Glass/Epoxy, compared to 1.4% for IM7/8552), thus being able to absorb more energy from the impactor. Secondly, the NCF architecture significantly enhances the delamination behaviour, due to the presence of stitches in the NCF fabric which diverts the propagation of crack. In addition, the fibrillar nature of Vectran fibres promotes extensive fibre bridging in between each ply.

Figure 7 presents a fractographic observation of S2-Glass/Epoxy composite under Mode I Interlaminar Fracture Toughness (IFT) testing, where the propagation of crack was diverted parallel with the direction of the stitch, thus requiring a higher

(a)   (b)

Fig. 7 Fractographic observation on the mode I behaviour of S2-glass/epoxy composite. **a** Crack growth morphology before a stitch, **b** crack growth morphology after the stitch. Notice the diversion of crack when propagating close to the stitch (Syed Abdullah 2019, 2021)

energy to further propagate the crack. Consequently, the delamination behaviour of the composite is significantly enhanced, resulting in a superior LVI performance.

8 Summary

While there may be many favourable attributes of composite material such is its high strength-to-weight ratio compared to conventional materials such as steel or aluminium, the susceptibility of composite towards impact loading is an inherent weakness which needs to be improved. The inherent weakness is mainly due to the small plastic component in the material, thus limiting its ability to a absorb a large amount of impact energy. This is especially true for classical brittle fibre-matrix system, such as CF/Epoxy, in addition to its relatively small tensile strain-to-failure. Under LVI, damage such as the BVID can increase the chances of catastrophic failure or sudden damage since no clear visual evidence of permanent indentation can be seen. It is quite possible for a composite structure to suffer fibre damage and massive delamination between plies, therefore more work is needed to detect the severity of damage (such as using a non-destructive testing approach). Therefore, it is important to improve the impact performance, by using an alternative approach such as hybridisation, where two (or more) types of reinforcements are included in a matrix material. Some of the reinforcements include polymer-based fibres such as Aramid and Vectran, which possess a large amount of plastic component in the material, thus being able to absorb more impact energy. In addition, hybridisation can also improve the compressive properties of polymer fibres, which is inherently poor, thus preventing it from being used as a structural component.

Perhaps more importantly is the need to understand the mechanics of failure due to a number of reasons. First, with a comprehensive understanding, engineers could exploit the full potential of the composite, without the need to over-design to ensure that the composite structure meets all safety requirements. This would result in a more cost-effective design, with minimal material consumption and fewer man-hours. Secondly, an accurate prediction of failure can be made, leading to a more reliable design since the mechanics of the composite are fully understood. Thirdly, physically based constitutive equations can be derived, which can be implemented as a material model for Finite Element Method (FEM) modelling. The material model can then be used to reduce the number of experimental tests required, leading to a reduction in the total design cost.

References

Abrate S (1991) Impact on laminated composite materials. Appl Mech Rev 44(4):155–190

Abrate S (1991) Impact on laminated composite materials: recent advances. Appl Mech Rev 47(11):517–544

Abrate S (1998) Imapct on composite structures. Cambridge University Press, Cambridge

AGY (2006) High strength glass fibres—technical datasheet. AGY

American Standard for Testing Materials (ASTM) (2017) ASTM D7137/D7137M–17, Standard test method for compressive residual strength properties of damaged polymer matrix composite plates. ASTM International, West Conshohocken

Argawal S, Singh KK, Sarkar PK (2014) Impact damage on fibre reinforced polymer matrix composite—a review. J Compos Mater 48(3):313–332

Arikan V, Sayman O (2015) Comparative study on repeated impact response of E-glass fibre reinforced polypropylene and epoxy matrix composites. Compos B Eng 83:1–6

Bandaru AK, Ahmad S, Bhatnagar N (2001) Impact behaviour of Aramid fibre/glass fibre hybrid composite: evaluation of a four-layer hybrid composites. J Mater Sci 36(9):2359–2367

Berthelot JM (2003) Transverse cracking and delamination in cross-ply glass-fibre and carbon-fibre reinforced plastic laminates: static and fatigue loading. Appl Mech Rev 56(2):111–147

Bouvet C, Castanie B, Bizeul M, Barrau JJ (2009) Low velocity impact modelling in laminate composite panels with discrete interface elements. Int J Solids Struct 46(14–15):2809–2821

Cantwell WJ, Morton J (1985) Detection of impact damage in CFRP laminates. Compos Struct 3(3–4):241–257

Cantwell WJ, Morton J (1989) Comparison of the low and high velocity impact. Composites 20(6):545–551

Cantwell WJ, Morton J (1989) Geometrical effects in the low velocity impact response of CFRP. Compos Struct 12(1):39–59

Cantwell WJ, Morton J (1991) The impact resistance of composite materials: a review. Composites 22(5):347–362

Caprino G, Lopresto V (2001) On the penetration energy for fibre-reinforced plastics under low-velocity impact conditions. Compos Sci Technol 61(1):65–73

Cowie JM (1991) Polymers: chemistry and physics of modern materials. Blackie and Son Limited, London

Crupi V, Epasto G, Guglielmino E (2016) Internal damage investigation of composites subjected to low-velocity impact. Exp Tech 40(2):555–568

Cuniff PM (1999a) Dimensionless parameters for optimization of textile-based body armor systems. In: International symposium of ballistics. San Antonio, Texas.

Cuniff, PM (1999b) Dimensionless parameters for optimization of textile-based body armour systems. In: 18th international symposium of ballistics. San Antonio, Texas

Davies GA, Olsson R (2004) Impact on composite structures. Aeronaut J 108(1089):541–563

Davies GA, Robinson P (1992) Predicting failure by debonding/delamination. In: AGARD 74th structures and materials meeting. debonding/delamination of composites. Neuilly sur Seine, France

de Freitas M, Silva A, Reis L (2000) Numerical evaluation of failure mechanisms on composite specimens subjected to impact loading. Compos Part B Eng 31(3):2000

Donald A, WIndle A, Hanna S (2006) Liquid crystal polymers as structural materials. Cambridge University Press, Cambridge

Dorey, G. (1987). Impact damage in composites—development, consequences, and prevention. In: Proceedings of the 6th international conference on composite materials (ICCM-6) and the 2nd European conference on composite materials (ECCM-2). London

Dorey G, Bishop SM, Curtis PT (1985) On the impact performance of carbon fibre laminates with epoxy and PEEK matrices. Compos Sci Technol 23(3):221–237

El-Zein M, Reifsnider K (1990) On the prediction of tensile strength after impact of composite laminates. J Compos Tech Res 12(3):147–154

Evci C, Gulgec M (2012) An experimental investigation on the impact response of composite materials. Int J Impact Eng 43:40–51

Garcia-Rodriguez SM, Costa J, Bardera A, Singery V, Trias D (2000) A 3D tomographic investigation to elucidate the low-velocity impact resistance, tolerance, and damage sequence of thin-non-crimp fabric laminates: effect of ply-thickness. Compos a Appl Sci Manuf 31(3):53–65

Godwin EW, Davies GA (1988) Impact behaviour of Thermoplastic composites. In: Composite materials technology. Springer, Berlin

Greenhalgh E (2009) Failure analysis and fractography of polymer composites. Woodhead Publishing, London

Hart KR, Chia PX, Sheridan LE, Wetzel ED, Sottos NR, White SR (2017) Comparison of compression-after-impact and flexure-after-impact protocols for 2D and 3D woven fiber-reinforced composites. Compos Appl Sci Manuf 101:471–479

Hazzard MK, Curtis PT, Iannucci L, Trask RS (2017) Effect of fibre orientation on the low velocity impact response of thin dyneema composite laminates. Int J Impact Eng 100:35–45

Heimbs S, Wagner T, Lozoya JT, Hoenisch B, Franke F (2018) Comparison of impact behaviour of glass, carbon, and dyneema composites. Proc Inst Mech Eng Part C J Mech Eng Sci

Hexcel (2018) Hextow IM7 carbon fibre—technical datasheet. Hexcel

Iannucci L (2018) Design of composite ballistic protection systems. Comprehensive composite materials II. Elsevier, London, pp 308–331

Laffan MJ, Pinho ST, Robinson P, McMillan AJ (2012) Translaminar fracture toughness of composite: a review. Polym Testing 31(3):481–489

Jih CH (1993) Prediction of delamination in composite laminates subjected to low velocity impact. J Compos Mater 27(7):684–701

Karthikeyan KR, Fleck NA, Wadley HN, Deshpande VS (2013) The effect of shear strength on the ballistic response of laminated composite plates. Euro J Mech Solids 42:35–53

Karthikeyan K, Kazemahvazi S, Russell BP (2016) Optimal fibre architecture of soft-matrix ballistic laminates. Int J Impact Eng 88:227–237

Katafiasz TJ, Iannucci L, Greenhalgh ES (2019) Development of a novel compact tension specimen to mitigate premature compression and buckling failure modes within fibre hybrid composites. Compos Struct 207:93–107

Khondker OA, Leong KH, Herszberg I, Hamada H (2005) Impact and compression-after-impact performance of weft-knitted glass textile composites. Compos a Appl Sci Manuf 35(5):636–648

Koh AC, Shim VP, Tan VB (2010) Dynamic behaviour of UHMWPE yarns and addressing impendance mismatch effects of specimen clamps. Int J Impact Eng 37(3):324–332

Kuraray America (2006) Vectran: liquid crystal technology—technical report. Kuraray

Liaw B, Delale F (2007) Hybrid carbon-glass fibre/toughened epoxy thick composite joints subjected to drop weight and ballistic impacts. US Army Research Office, New York

Liu D (1988) Impact induced delamination—a view of bending stiffness mismatching. J Compos Mater 22(7):674–692

Miao H, Wu Z, Ying Z, Hu X (2019) The numerical and experimental investigation on low-velocity impact response of composite panels: Effect of fabric architecture. Compos Struct 227:1–11

Mitrevski T, Marshall IH, Thomson R (2006) The influence of impactor shape on the damage of composite laminates. Compos Struct 76(1–2):116–122

Mitrevski T, Marshall IH, Thomson R, Jones R, Whittingham B (2015) The effect of impactor shape on the impact response of composite laminates. Compos Struct 67(2):139–148

Naik and Shrirao, 2004. Naik NK, Shrirao PS (2004) Composite structures under ballistic impact. Compos Struct 66(1–4):579–590

Naik NK, Shrirao P, Reddy BC (2005) Ballistic impact behaviour of woven fabric composites: parametric studies. Mater Sci Eng 412:104–106

Olsson R (2001) Analytical prediction of large mass impact damage in composite laminates. Compos Appl Sci Manuf 32(9):1207–1215

Olsson R (2010) Analytical model for delamination growth during small mass impact on plates. Int J Solids Struct 47(21):2884–2892

Olsson R, Donadon MV, Falzon BG (2006) Delamination threshold load for dynamic impact on plates. Int J Solids Struct 43(10):3124–3141

Pandya KS, Pothnis JR, Ravikumar G, Naik NK (2008) Stress wave attenuation in composites during ballistic impact. Polym Testing 31(2):261–266

Park R, Jang J (2001) Impact behaviour of Aramid fibre/glassfibre hybrid composite: evaluation of a four-layer hybrid composites. J Mater Sci 36(9):23592367

Park R, Jang J (2004) Impact behaviour of aramid fibre/glass fibre hybrid composites the effect of stacking sequence. Polym Compos 22(1):80–89

Puck S, Schurman H (1998) Failure analysis of FRP lamiants by means of physically based phenomenological models. Compos Sci Technol 58(7):1045–1067

Olsson R, Donadon MV, Falzon BG (2006) Delamination threshold load for dynamic impact on plates. Int J Solids Struct 43(10):3124–3141

Reddy TS, Reddy PR, Madhu V (2017) Response of E-glass/epoxy and dyneema composite laminates subjected to low and high velocity impact. Procedia Eng 173:278–285

Richardson MO, Wisheart WJ (1996) Review of low-velocity impact properties of composite materials. Compos Appl Sci Manuf 27(12):1123–1131

Robinson P, Davies GA (1992) Impactor mass and specimen geometry effects in low velocity impact of laminated composites. Int J Impact Eng 12(2):189–207

Saleh MN, El-Dessouky HM, Saeedifar M, Freitas ST, Scaife RJ, Zarouchas D (2019) Compression after multiple low velocity impacts of NCF, 2D and 3D woven composites. Compos Appl Sci Manuf 125:1–12

Sanchez-Saez S, Barbero EZ, Navarro C (2005) Compression after impact of thin composite laminates. Compos Sci Technol 65(13):1911–1919

Schoeppner GA, Abrate S (2000) Delamination threshold loads for low velocity impact on composite laminates. Compos Part A Appl Sci Manuf 31(9):903–915

Schoeppner GA, Abrate S (2000) Delmination threshold loads for low velocity impact on composite laminates. Compos Appl Sci Manuf 31(9):903–915

Shaktivesh NS, Sesha-Kumar CV, Naik NK (2013) Ballistic impact performance of composite targets. Mater Des 51:833–846

Sharma AP, Khan SH, Velmurugan R (2019) Effect of through thickness separation of fiber orientation on low velocity impact response of thin composite laminates. Heliyon 5(10):1–14

Shyr T-W, Pan YH (2003) Impact resistance and damage characteristics of composite laminates. Compos Struct 62(2):193–203

Sikarwar RS, Velmurugan R, Gupta NK (2017) Effect of velocity and fibres on impact performance of composite laminates—analytical and experimental approach. Int J Crashworthiness 22(6):589–601

Silberschmidt VV (2016) Dynamic deformation, damage, and fracture in composite materials and structures. Woodhead Publishing, Loughborough

Singh V (2018) Literature survey of strain rate effects on composites—technical Report. Swerea SICOMP, Sweden

Sjoblom PO, Hartness TJ (1988) On low-velocity impact testing of composite materials. J Compos Mater 22(1):30–52

Sonnenfeld C, Mendil-Jakani H, Agogue R, Nunez P, Beauchene P (2017) Thermoplastic/thermoset multilayer composites: a way to improve the impact damage tolerance of thermosetting resin matrix composites. Compos Strucut 171:298–305

Syed Abdullah SIB (2019) The impact behaviour of high performance fibre composites, Ph.D. thesis. Imperial College London, London

Syed Abdullah SIB, Iannucci L, Greenhalgh, ES (2018) On the translaminar fracture toughness of Vectran/epoxy composite material. Compos Struct 202:566–577

Syed Abdullah SIB, Iannucci L, Greenhalgh ES, Ahmad Z (2021) The impact performance of Vectran/Epoxy composite laminates with a novel non-crimp fabric architecture. Compos Struct 265:1–17

Tan KT, Watanabe N, Iwahori Y, Ishikawa T (2012) Effect of stitch density and stitch thread thickness on compression after impact strength and response of stitched composites. Compos Sci Technol 72(5):587–598

Tan VB, Zeng XS, Shim VP (2010) Characterization and constitutive modelling of Aramid fibres at high strain rate. Int J Impact Eng 37(3):324–332

Vaidya UK (2011) Impact response of laminated and sandwich composites. Impact engineering of composite structures. Springer, Udine, pp 97–191

Vieille B, Casado VM, Bouvet C (2013) About the impact behavior of woven-ply carbon fiber-reinforced thermoplastic and thermosetting composites: a comparative study. Compos Struct 101:9–21

Wang Y, Xia YM (1998) The effects of strain rate on the mechanical behaviour of Kevlar fibre bundle: an experimental and theoretical study. Compos a Appl Sci Manuf 29(11):1411–1415

Williams KV, Vaziri R, Poursartip A (2003) A physically based continuum damage mechanics model for thin laminated composite structures. Int J Solids Struct 40(9):2267–2300

Wisnom MR (2012) The role of delamination in failure of fibre-reinforced composites. Philos Trans Roy Soc 370(1965):1850–1870

Zhou G (2003) Prediction of impact damage thresholds of glass fibre reinforced laminates. Compos Struct 62(2):193–203

Zhou J, Liao B, Shi Y, Zuo Y, Tuo H, Jia L (2019) Low-velocity impact behaviour and residual tensile strength of CFRP laminates. Compos B Eng 161:300–313

Experimental Analysis of High Velocity Impact Properties of Composite Materials for Ballistic Applications

Suhas Yeshwant Nayak, Rashmi Samant, B. Satish Shenoy, M. T. H. Sultan, and Chandrakant R. Kini

Abstract This chapter presents a review of the use of composites as personnel protective armours. It lays out the various standards and specifications that are used in evaluating the effectiveness of an armour for personnel protection. The NIJ standard which is popularly used to evaluate the armours is thoroughly discussed along with common terminologies associated with the same. The study also explores the various testing equipment and ammunition used for testing from shapes to materials and their impact on the armour panels. It is essential to follow the standards meticulously to ensure safety and success of any testing. Composites are gaining increased prominence in modern day warfare and has evolved from use of metals since the days of the first and second world wars.

Keywords Personnel protective armour · Ballistic impact · Composite materials · Standard testing

S. Y. Nayak · R. Samant
Department of Mechanical and Manufacturing Engineering, Manipal Institute of Technology, Manipal Academy of Higher Education, Manipal, Karnataka 576104, India

B. S. Shenoy (✉) · C. R. Kini
Department of Aeronautical and Automobile Engineering, Manipal Institute of Technology, Manipal Academy of Higher Education, Manipal, Karnataka 576104, India
e-mail: satish.shenoy@manipal.edu

M. T. H. Sultan
Laboratory of Biocomposite Technology, Institute of Tropical Forestry and Forest Products (INTROP), Universiti Putra Malaysia, 43400 UPM Serdang, Selangor, Malaysia

Department of Aerospace Engineering, Faculty of Engineering, Universiti Putra Malaysia, 43400 UPM Serdang, Selangor, Malaysia

M. T. H. Sultan
e-mail: thariq@upm.edu.my

© Springer Nature Singapore Pte Ltd. 2021
M. T. H. Sultan et al. (eds.), *Impact Studies of Composite Materials*, Composites Science and Technology, https://doi.org/10.1007/978-981-16-1323-4_2

1 Introduction to Ballistics

Ballistic is a division of applied physics which deals with the motion of projectiles propelled by different energy sources like solid, liquid, gas, electrical, electromagnetic and laser sources. This field can be further divided into four categories. They are interior ballistics, intermediate ballistics, exterior ballistics and terminal ballistics. Interior ballistics is concerned with combustion of material and the propagation of gas in the gun/rocket whereas intermediate ballistics deals with behaviour of a projectile while leaving the barrel/launcher. Exterior ballistics is study of motion of a missile/projectile/rocket after the launch from a platform of a muzzle of a weapon. The fourth category, terminal ballistics is study of effects of projectiles after they have reached the target (Bilisik 2017). Velocities greater than 50 m/s but less than 1000 m/s are considered as high velocity. Any velocity below 50 m/s is termed as low velocity while above 1000 m/s is referred to hyper velocity (Ismail et al. 2019). Projectiles travelling at such a velocity has potential to cause excessive damage in spite of the fact they are light in weight. Some of the examples of ballistic impact include bird hitting airplanes in their flight, hailstorm, debris hit causing damage to automobile structure, military applications like firing bullets, flying sharpnels from a bomb blast, etc. (Salman et al. 2015). Thus it is imperative to develop materials which can provide protection against damage caused due to ballistic impact. The material should be efficient enough to provide complete protection but at the same time should be cost effective, light in weight, easy to fabricate. Continual research and development has resulted into shift from materials like manganese to polymer composite materials. The focus of this chapter will be on development of composite materials, mechanism of composite materials in absorbing impact energy, standards used, different testing methods, terminologies in ballistic testing and failure modes.

2 Composite Materials as Anti-Ballistic Materials

During the Second World War, armours were mainly made with heavy manganese and steel which restricted personnel movement and used to overheat. These disadvantages eventually led to the use of manganese steel sheets within multi-layered nylon. Further development led to all-nylon armour without any metal. Fibre reinforced plastic armours could be used over a range of temperature but they failed in protection from rifle projectiles. In 1965 aramid fibre was developed with light weight and, high stiffness and strength. Several high-performance fibres developed were, Twaron® and Kevlar®, Dyneema® and Spectra® from Ultra High Molecular Weight Polyethylene, and Polybenzoxazole (PBO) fibre (Rosenberg and Dekel 2016; Zaera 2011).

Later, Multi-layered Ballistic Armour System (MBAS) also known as dual-hardness armours with ductile backing material made up of high performance fibre and brittle hard front face for absorbing kinetic energy were developed. For the front face, materials like steel, titanium, aluminium, aluminium nitride, titanium diboride, silicon carbide, zirconium oxide and boron carbide were used (Rosenberg and Dekel 2016). Presently, along with synthetic fibres, natural fibre reinforcements are also being explored as they are inexpensive and hybridized polymer composites have reported superior ballistic performance (Salman et al. 2015; Zakikhani et al. 2016). Any composite material consists of two distinct phases, matrix which is a continuous phase and reinforcement which is discontinuous phase. Based on matrix, composite materials can be broadly classified into three categories viz. polymer matrix composites, metal matrix composites and ceramic matrix composites. Other than fibres, reinforcements come in forms, like particulates, flakes, and whiskers (Schwartz 1984).

3   Mechanism of Shock Absorption During Ballistic Impact

Any projectile travelling will possess kinetic energy. As the projectile impacts the armour this energy acts over a very small area and allows the projectile to perforate through the materials. In general armour absorbs this kinetic energy and spreads it over a large area thereby making it difficult for the projectile to punch through. Modern day armours make use of woven fabric in armour systems. The yarns in theses woven fabrics are known to have high specific strength and modulus (Mostafa et al. 2016). High modulus of yarns results in dissipation of the energy along its length. As the dissipated energy meets a junction in the woven fabric, it gets divided by a number of possible mechanisms. It may continue along the yarn, it may get reflected back or it may travel along the crossing yarn. This dissipation and division of energy takes place at several such junctions and in several layers in the armour until the projectile has lost sufficient amount of energy that it cannot further penetrate into the material. As the projectile impacts into the first layer, shearing of the layer takes place and also absorbs some amount of energy. Figure 1 shows the distribution of energy along a yarn of the fibre.

Fig. 1 Mechanism of defeating the projectile by distribution of kinetic energy (Cooper and Gotts 2005)

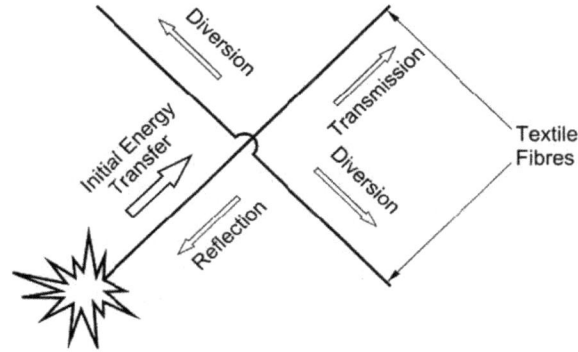

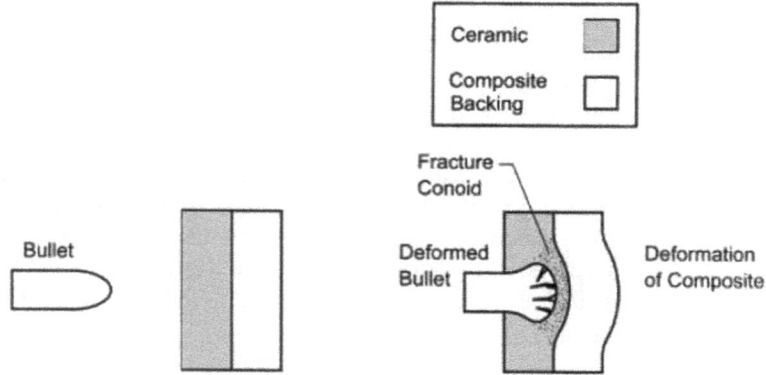

Fig. 2 Mechanism of defeating the projectile armours with hard front face (Cooper and Gotts 2005)

Absorption of energy due to shearing of the fabric layer is another mechanism by which the projectile is defeated in its travel. Mostly all the types of armours nowadays uses a hard front face which is responsible for distorting the projectile before the backing composites spreads the energy over a larger area. Figure 2 depicts the mechanism by which the projectile is defeated with the combination of hard front face and a backing composite layer (Cooper and Gotts 2005).

4 Ballistic Testing

4.1 Standards of Ballistic Testing

Standards are imperative to all forms of studies conducted. They provide basis for comparison, deduction and improvising existing or new methods/materials for varied applications. They serve as a set of universal rules for testing and maintaining uniformity in recording results.

The various standards used for ballistic testing are US National Institute of Justice, European Committee for Standardization, Joint Technical Committee MS/43 (Australia and New Zealand), State Standardization Committee of Russian Federation, North Atlantic Treaty Organization and US Department of Defence. These standards vary from each other in terms of scope of application. Various standards cater specifically and efficiently to a class of materials such as ballistic helmets, armoured vehicles whereas others cover all ballistic materials. Another major point of difference between the standards is the level of threat. This is accounted in terms of gun calibre and ammunition employed (Zaera 2011). The US National Institute of Justice is the most widely used and recognized standards among the many.

Table 1 Levels of protection and its details (Samuels 2000)

Level of protection	Size of bullet	Mass of bullet	Velocity
IIA	9 mm FMG RN	8 g	373 m/s ± 9.1 m/s
	0.40 S&W	11.7 g	352 m/s ± 9.1 m/s
II	9 mm FMG RN	8 g	398 m/s ± 9.1 m/s
	0.357 Magnum	10.2 g	436 m/s ± 9.1 m/s
IIIA	0.357 SIG	8.1 g	448 m/s ± 9.1 m/s
	0.44 Magnum	15.6	436 m/s ± 9.1 m/s
III	7.62 mm FMJ	9.6	847 m/s ± 9.1 m/s
IV	0.30 caliber AP	10.8 g	878 m/s ± 9.1 m/s

For body armours to be tested for ballistic resistance, the NIJ standard—0101.06 is the most imperative. The scope of this standard is to ascertain the limits of performance of personnel protective gear and test methods for ballistic applications against gunfire. Table 1 has all the details of the various levels of protection as put down by NIJ standards.

4.2 Important Terminologies

There are a number of terminologies used to denote various aspects of ballistic testing with some of them being more repetitive and prominent than others. It is discussed in detail over Table 2.

4.3 Interpretation of the Test Results

Analysis and interpretation of results conclude any process and establish the success rate of the tests conducted. Various methods and standards have been developed to evaluate various categories of testing and analysis. The armour specimens are evaluated based on the back face signature and ballistic limit values ascertained from the testing.

Impact velocity meters and residual velocity meters are used to record the values of ballistic limit for the projectiles. Ballistic limit varies with shape of the projectiles. A projectile is expected not to penetrate the composite panel below the ballistic limit. Hence suitability of composite panels for ballistic applications is decided based on

Table 2 Important terminologies (Samuels 2000)

Terminology	Definition
Angle of incidence	The angle between the line of bullet strike and the perpendicular to the surface of backing material (Fig. 3) at the point of incidence is termed as angle of incidence
Armour carrier	A non-ballistic resistant material that is employed to secure the armour material to the body of the user
Armour conditioning	Values of mechanical an environmental parameters of the armour before testing. It includes humidity, temperature and mechanical damage
Back face signature	The highest depth of indentation caused by a bullet that does not pass through the armour being tested is termed as back face signature
Baseline ballistic limit	It is the ballistic limit value derived experimentally for a new ballistic armour panel
Backing material	A layer of oil based clay stationed in close contact with the armour being tested is called backing material
Backing material Fixture	It is a rigid structure shaped like a box with a detachable back which houses backing material. This detachable back is employed during perforation-backface testing and not during V50 testing
Ballistic limit	For a particular type of bullet the velocity at which the bullet is expected to perforate the armour panel at a probability of 50% is termed as Ballistic Limit. The Ballistic Limit is also referred to as V50
Compliance test group	A batch of armour panels turned in for testing as per a particular standard
Dew point	The temperature of an air parcel which is required to be cooled to, keeping constant barometric pressure in order to condense the water vapour to water (dew)
Fair hit	Refers to the impact created by the bullet on the composite armour panel subjected to standard velocity requirement and shot spacing
Full metal jacketed bullet (FMJ)	Refers to a lead bullet coated with copper alloy on all surfaces except the base. The alloy consists of 90% copper and 10% Zinc

comparing the general expected velocity of real ammunition to the V_{50} (Ballistic Limit).

Ballistic limit is found to be independent of the thickness of the projectile but is dependent on the sharpness of the projectile. It varies inversely with the sharpness of the projectile (discussed in Sect. 4.5) used with highest numerical value recorded for blunt projectiles (Ansari and Chakrabarti 2017). Ballistic limit increases with increase in thickness of the composite panels (VanderKlok et al. 2018). Shear strength of composites play a major role in deciding the ballistic limit since the major means of failure involve shear snapping. The modes of failure are discussed in the next

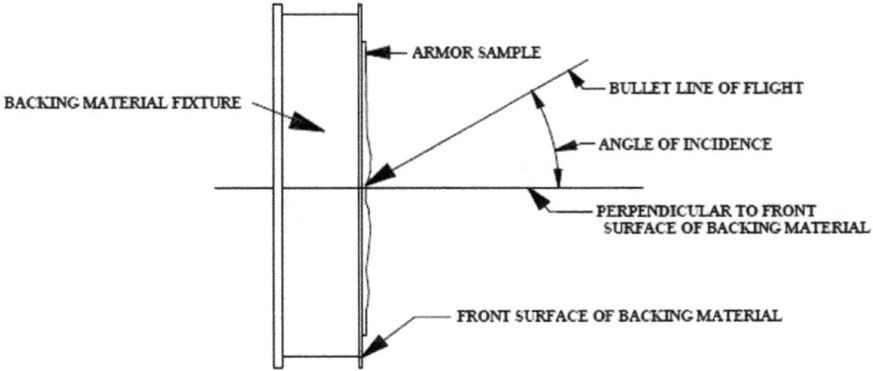

Fig. 3 Angle of incidence (Samuels 2000)

section (Sect. 5), The decrease in shear strength causes an almost doubling increase in the values of ballistic limits for composite panels (Wang et al. 2017).

Based on the type of testing being done as per the values prescribed by the standards, the ballistic limit is evaluated. The numerical value is ascertained using standard methods. The first one involves the velocity time history where V_{50} is assigned to the maximum impact velocity which stops the projectile. In the second method (specified by US MIL-STD-662E), an average of the lower range of velocities displaying full perforation is taken along with the higher range of velocities having only partial perforations. The range for the higher band is observantly very small (Bandaru et al. 2016).

If a NIJ Standard-0106.01 9 mm FMJ (full metal jacket), type IIIA testing is carried out on a Kevlar composite for probable ballistic applications. While testing an impact velocity of 380 and 400 m/s brings in residual velocities of 68 and 131 m/s. The value of residual velocity begins to approach 0 for an impact velocity of 376 m/s. At this point a number of tests are carried out and an average of impact velocities with 0 residual velocities is arrived at as the ballistic limit of the composite panel (Bandaru et al. 2016).

The composite panel when mounted for testing is generally covered with an additional layer of oily clay to record the impact of the projectile on the laminate. The back face material is then evaluated to obtain the back face signature of a particular projectile. The depth of indentation on the back face material is recorded every time and the material is passed for application if all values fall below 44 mm. The depth is measured with a Vernier calipers as showed in Fig. 4. This is the value above which it is prescribed to be lethal to human beings as per the NIJ standard 0101.04 (Fabio et al. 2017).

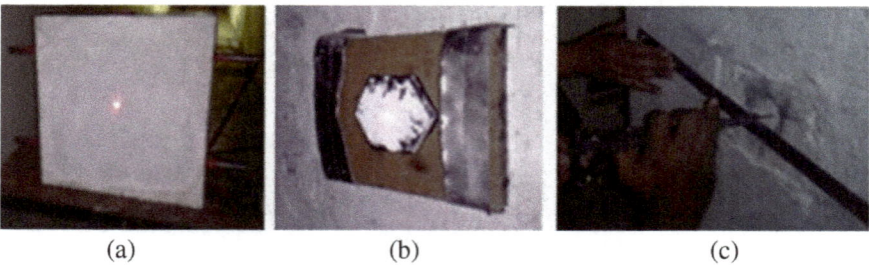

Fig. 4 **a** Clay backing material, **b** testing panel mounted on clay witness, **c** measurement of back face signature (Fabio et al. 2017)

4.4 Testing Methods

Testing is essential in determining the success of fabrication and suitability of the fabricated composite material for ballistic applications. Since human lives are on line during warfare and combat involving ballistics, a range of carefully curated tests have to be carried out and a range of values iterated before the composite can be approved for application. Ballistic armours are tested using a wide variety of methods in Universities, Independent Research facilities and defence/ military organisations in accordance with set standards of testing parameters.

In universities and research institutes, the most popular testing methods include the use of single stage gas guns, two-stage gas guns and powder barrel guns. The gas guns use pressurized gases like helium, nitrogen etc. to propel ammunition towards the armour panel being tested. The velocity of the ammunition fired is controlled by varying the pressure of the gas. The setup consists of a gas supply cylinder, a high pressure cylinder, barrel and muzzle in addition to an exhaust and a valve. Figure 5 is a schematic representation of the test setup for ballistic armours. The

Fig. 5 Typical arrangement of test setup (Luz et al. 2017)

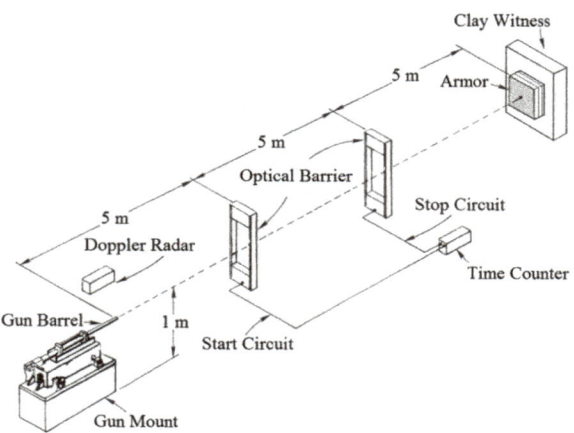

Fig. 6 Field test set-up
(Purushothaman et al. 2013)

Gun is mounted to face the armour which is fastened to a clay witness which acts as the backing material. It records the back face signature of the projectile in case of non-perforation.

Military establishments make use of various caliber of guns such as an AK-47 in actual application during warfare or combat. They make use of Field Test Set-up as shown in Fig. 6.

Military operations are highly unpredictable and involve close combat and stabbing at times. Hence low velocity impact tests are mostly carried out along with the general high velocity impact tests using guns. Low velocity indenters are used for testing low speed penetration ballistic applications. Personnel protective equipment is tested by using instrumented drop weight testers (Reddy et al. 2017b). The indenters are dropped from a height and strain gauges are used on the armour panels to record the impact energy. Various types and shapes of knives are used as indenters in testing. Double-edged steel knives are a popular choice.

4.5 Types of Projectiles

A variety of Projectiles are used to simulate an actual firing sequence from the fields. Universities and research establishments use a range of projectiles for the same. They give the researchers a real-time experience to further studies on materials with dangerous field applications. Various materials and shapes of projectiles are employed for ballistic testing of armours.

Steel is a popular choice of material for making ammunition. Grades like steel 1020, hardened steel 4340 etc. are employed (Li et al. 2017; Vinson and Walker 1997). Apart from steel, projectiles are also made from brass, silica, aluminium, copper, and lead. (Meng et al. 2017; Li et al. 2017; Shockey et al. 1975).

Hemispherical, Flat, Ogival, Conical as shown in Fig. 7 are some of the most common shapes for projectiles. They are characterised by their diameter, height and specific weight. The velocity of the projectiles is adjusted to the specific weight of

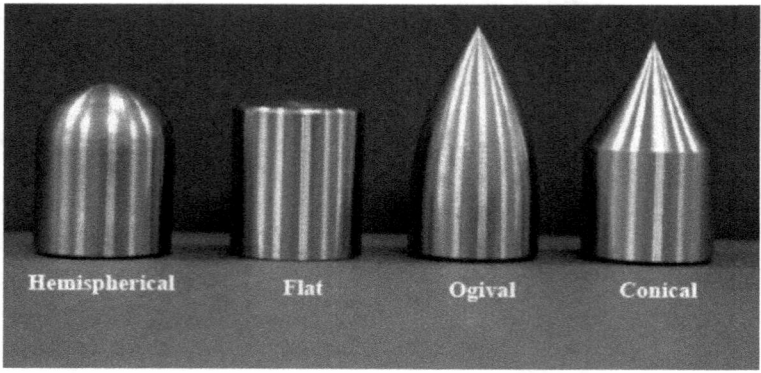

Fig. 7 Types of projectiles used (Tan et al. 2003)

each projectile. Elongated and sharper shapes are known to have lower ballistic limits but there are exceptions to this as well (Tan et al. 2003).

5   Modes of Failure in Composite Armour Panels

Studying the modes of failure is a major step in deciding the efficiency of existing fabrication methods and strength of the fabricated composite panels. Failure is synonymous with extent of bonding between matrix and fibres, rate of energy dissipation and the overall scope for improving current methods and materials. Delamination, matrix cracking, de-bonding, fibre breakage and shear plugging are major means of failure during ballistic testing of composite panels. The failure might be attributed to one of the above phenomenon or a combination of a few of them. This is caused due to high velocity impact and the following interactions between the projectiles and the composite material. It is a function of shape of projectile, mass of projectile, velocity of projectile and distance from which it is fired.

Fibres snap by the sheer force exerted by the sharp edges of the projectile. This is mainly caused by the tension at the back of the composite panel. Fibre breakage as seen in Fig. 8 happens close to the point of impact of bullet and is scarcely detected towards the outer edges of the panel. The panel in Fig. 8 is a glass–epoxy composite. The white arrows in the figure point towards of the area of damage. The image is a lateral cut section of the panel close to the area of damage (Ávila et al. 2011).

Delamination is attributed to fracturing between layers of a panel. It is a macroscopic phenomenon and detectable by visual inspection. Figure 9 shows one such example. Delamination is a major mode of failure and happens due to collective action of micro fractures propagating through the material as an impulse. Impulsive force is directly proportional to the velocity of impact and hence the permanent deformations increase with increase in velocity (Li et al. 2017). Delamination is a function

Fig. 8 Fibre breakage (Ávila et al. 2011)

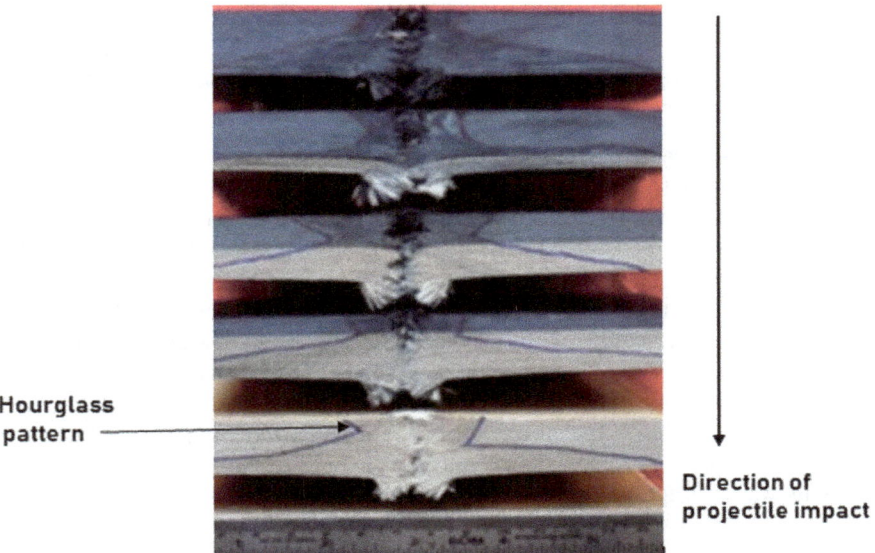

Fig. 9 Overview of delamination in composite panels (Reddy et al. 2017a)

of laminate thickness. An hourglass pattern is observed on the cross-section of panels at the point of impact. The base of the hourglass pattern expands with increase in the thickness pointing to increased extent of delamination in the composite panels (Reddy et al. 2017a).

De-bonding as shown in Fig. 10 is another macroscopic failure characterized by the split of fibres from the matrix. It is prevalent in composites where the resin and fibre have weak interfacial bonds (Benzait and Trabzon 2018). The arrows in Fig. 10 point towards debonding across the panel. Matrix cracking is caused by the impulse force and leads to de-bonding in many of the cases. Miniature fractures propagate through the matrix of the panel from the point of impact.

Shear plugging is caused by compressive load under the projectile. A plug is formed beneath the volume occupied by the projectile in the composite material. It is

Fig. 10 De-bonding
between the fibre and matrix
(Ávila et al. 2011)

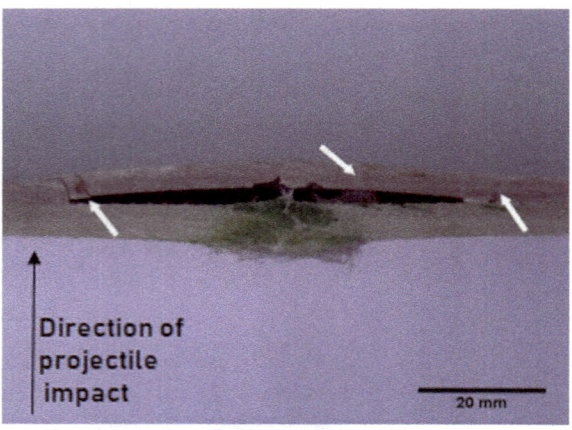

attributed to displacement of composite volume suspended by the fibres at the edges of the plug. It causes macroscopic failure. Figure 11 depicts a simulation of failure by shear plugging as highlighted by the circular region. The panel is simulation of ceramic fabric reinforced metal matrix composite armour (McWilliams et al. 2016).

There are many mechanisms of interaction between the projectile and composite armours as discussed above. A detailed view of the same can be obtained by studying the interaction of projectiles with stacked up fabrics. Many researchers have lead studies by repeating the tests conducted on composite panels with stacks of woven fibre mats to better understand minute phenomenon that lead to major failures.

The first interaction between the projectile and the fibres is attributed to three probable mechanisms as depicted in Fig. 12. In the first scenario (a), the ammunition forces its way through the material by sheer momentum and velocity. It leaves an

Fig. 11 Failure by shear
plugging (McWilliams et al.
2016)

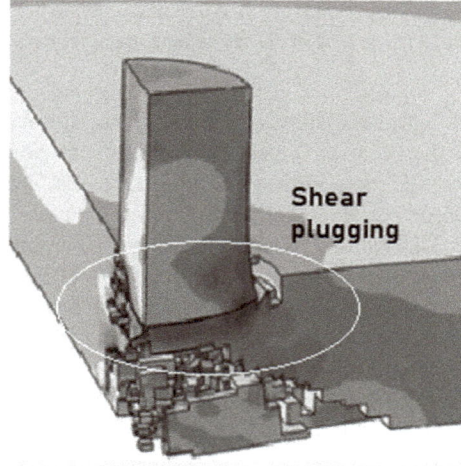

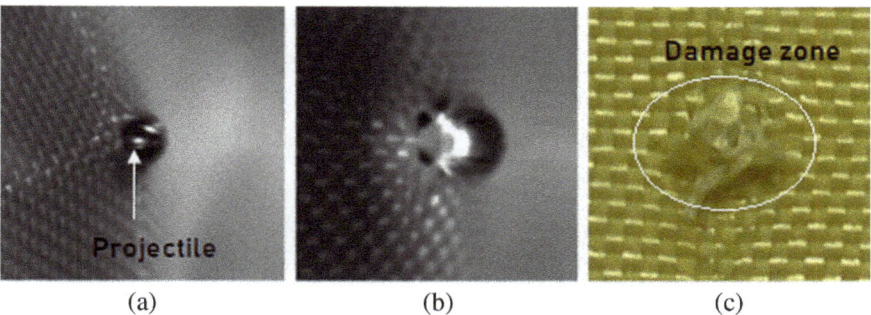

(a)   (b)   (c)

Fig. 12 **a** A hemispherical projectile windowing through the weave, **b** a yarn spreading and sliding out of the way of a hemispherical projectile, and **c** the cutting of some fibres/yarns around the nose of a sharp projectile (Cline et al. 2020)

impression at the point of entry. The second scenario (b) arises when the sharp tip of the projectile slips through the pore of the woven material and enlarges the pore into a rupture as it progresses through the material. The third mechanism (c) involves snapping of yarns and fibres due to the sharp tip of the projectile creating a pathway for the same to enter the ballistic panel (Cline et al. 2020).

The stacked panel fails due to a number of reasons. Yarn rupturing, fibre splitting, fibrillation, friction and bowing are major mechanisms. During the rupture of the fibre stack, the fibres making up the armour snap in a disorderly fashion to cause a depression in the surface of the panel as shown in Fig. 13. Rupturing is majorly attributed to breaking of bonds in the minute scale. These bonds are generally covalent in nature. Disorderly yarn pullout is the indication of failure by rupture. It is generally caused by blunt force trauma to the panels. Figure 13 shows failure of fibre panes caused by (i) Hemispherical; (ii) Flat head; (iii) Ogival head and (iv) Conical head projectiles respectively.

Fibrillation is the process of snapping of fibres into two. This happens due to breakage of secondary bonds, mainly hydrogen bonding between the molecules of the fibre due to the high velocity impact caused by the projectile and the angle of penetration. Friction is another cause of failure. It happens due to the abrasion between the bullet surface and the armour material at high velocity.

A circular or oval disturbance is the signature of failure by bowing. The failure can be seen propagating through the rest of the panel in lines like fault lines from the site of a crack. It is a prevalent mode of failure in ballistic panels with orthogonally woven yarns. It's majorly attributed to the dislocation of yarns in perpendicular direction due to the incident wave of force accompanying the bullet. Figure 14 shows an example of failure seen due to bowing due to (i) Hemispherical; (ii) Flat head; (iii) Ogival head and (iv) Conical head shaped projectiles (Tan et al. 2003).

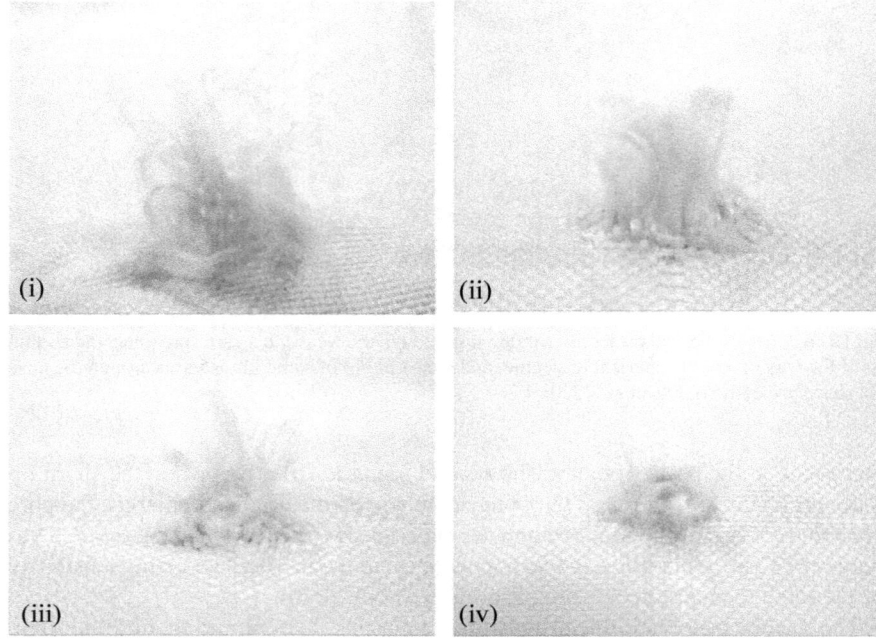

Fig. 13 Representation of failure due to rupture of yarns with impact of **i** Hemispherical; **ii** Flat head; **iii** Ogival head, **iv** Conical head (Tan et al. 2003)

6 Conclusion

This section comprehensively introduces the aspect of standardization in ballistic testing for personnel protective armor and safety vests employed in warfare. It explores the various shapes and materials used to make the projectiles and gives an insight to the steps of testing involved. Gas gun are most often employed for testing. The tested samples are mostly evaluated with respect to the ballistic limits and back face signatures and classified into classes of a standard. The most common modes of failure include debonding, matrix cracking and delamination. This process of standardized testing has meticulously advanced research and brought a sense of uniformity to the data and work going on around the world in the field of protective materials.

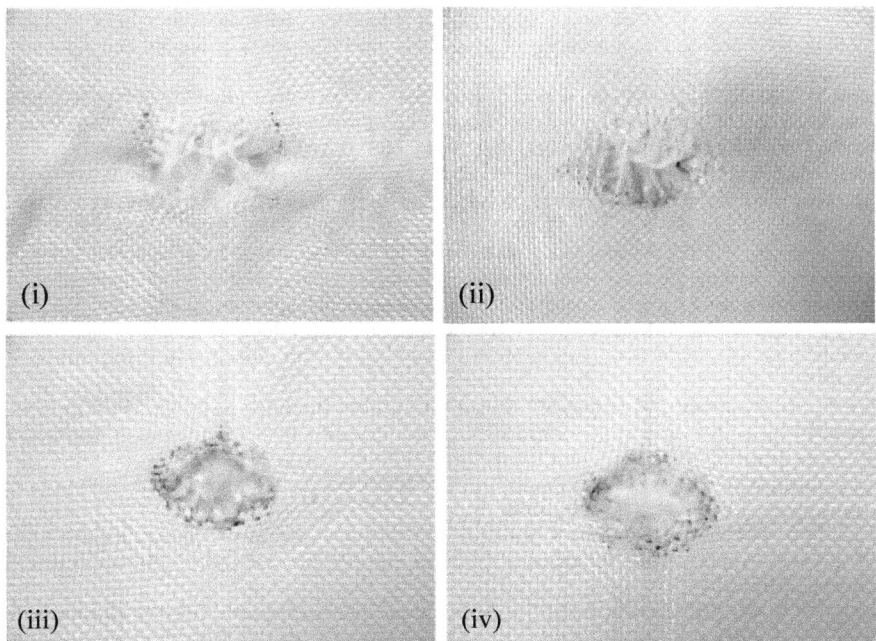

Fig. 14 Representation of failure due to bowing of yarns with impact of **i** hemispherical; **ii** flat head; **iii** ogival head, **iv** conical head (Tan et al. 2003)

References

Ansari MM, Chakrabarti A (2017) Influence of projectile nose shape and incidence angle on the ballistic perforation of laminated glass fiber composite plate. Compos Sci Technol 142:107–116. https://doi.org/10.1016/j.compscitech.2016.12.033

Ávila AF, Neto AS, Nascimento Junior H (2011) Hybrid nanocomposites for mid-range ballistic protection. Int J Impact Eng 38:669–675. https://doi.org/10.1016/j.ijimpeng.2011.03.002

Bandaru AK, Chavan VV, Ahmad S et al (2016) Ballistic impact response of Kevlar® reinforced thermoplastic composite armors. Int J Impact Eng 89:1–13. https://doi.org/10.1016/j.ijimpeng.2015.10.014

Benzait Z, Trabzon L (2018) A review of recent research on materials used in polymer–matrix composites for body armor application. J Compos Mater 52:3241–3263. https://doi.org/10.1177/0021998318764002

Bilisik K (2017) Two-dimensional (2D) fabrics and three-dimensional (3D) preforms for ballistic and stabbing protection: a review. Text Res J 87:2275–2304. https://doi.org/10.1177/0040517516669075

Cline J, Moy P, Harris D et al (2020) Ballistic response of woven Kevlar fabric as a function of projectile sharpness. Conf Proc Soc Exp Mech Ser 1:13–16. https://doi.org/10.1007/978-3-030-30021-0_3

Cooper G, Gotts P (2005) Ballistic protection. Ballistic trauma. Springer, London, pp 67–90

da Luz FS, Monteiro SN, Lima ES, Lima Júnior ÉP (2017) Ballistic application of coir fiber reinforced epoxy composite in multilayered armor. Mater Res 20:23–28. https://doi.org/10.1590/1980-5373-mr-2016-0951

Fabio L, Nascimento C, Henrique L, Louro L (2017) Ballistic performance in multilayer armor with epoxy composite reinforced with malva fibers. In: Proceedings of the 3rd Pan American materials congress, pp 331–338

Ismail MF, Sultan MTH, Hamdan A et al (2019) Low velocity impact behaviour and post-impact characteristics of kenaf/glass hybrid composites with various weight ratios. J Mater Res Technol 8:2662–2673. https://doi.org/10.1016/j.jmrt.2019.04.005

Li X, Nia AB, Ma X et al (2017) Dynamic response of Kevlar®29/epoxy laminates under projectile impact-experimental investigation. Mech Adv Mater Struct 24:114–121. https://doi.org/10.1080/15376494.2015.1107670

McWilliams BA, Yu JH, Pankow M (2016) dynamic failure mechanisms in woven ceramic fabric reinforced metal matrix composites during ballistic impact. In: Residual stress, thermomechanics and infrared imaging, hybrid techniques and inverse problems. Springer, pp 155–159

Meng Z, Singh A, Qin X, Keten S (2017) Reduced ballistic limit velocity of graphene membranes due to cone wave reflection. Extrem Mech Lett 15:70–77. https://doi.org/10.1016/j.eml.2017.06.001

Mostafa NH, Ismarrubie ZN, Sapuan SM, Sultan MTH (2016) The influence of equi-biaxially fabric prestressing on the flexural performance of woven E-glass/polyester-reinforced composites. J Compos Mater 50:3385–3393. https://doi.org/10.1177/0021998315620478

Purushothaman A, Coimbatore G, Ramkumar SS (2013) Soft body armor for law enforcement applications. J Eng Fiber Fabr 8:97–103

Reddy PRS, Reddy TS, Mogulanna K et al (2017a) Ballistic impact studies on carbon and E-glass fibre based hybrid composite laminates. Procedia Eng 173:293–298. https://doi.org/10.1016/j.proeng.2016.12.017

Reddy TS, Reddy PRS, Madhu V (2017b) Response of E-glass/epoxy and Dyneema® composite laminates subjected to low and high velocity impact. Procedia Eng 173:278–285. https://doi.org/10.1016/j.proeng.2016.12.014

Rosenberg Z, Dekel E (2016) Terminal ballistics, 2nd edn

Salman SD, Leman Z, Sultan MTH et al (2015) Kenaf/synthetic and Kevlar®/cellulosic fiber-reinforced hybrid composites: a review. BioResources 10:8580–8603

Samuels JE (2000) NIJ Standard–0101.04 (Ballistic Resistance of Personal Body Armor). 67

Schwartz MM (1984) Composite materials handbook. McGraw Hill, New York

Shockey DA, Curran DR, De Carli PS (1975) Damage in steel plates from hypervelocity impact. I. Physical changes and effects of projectile material. J Appl Phys 46(9):3766–3775

Tan VBC, Lim CT, Cheong CH (2003) Perforation of high-strength fabric by projectiles of different geometry. Int J Impact Eng 28:207–222. https://doi.org/10.1016/S0734-743X(02)00055-6

VanderKlok A, Stamm A, Dorer J et al (2018) An experimental investigation into the high velocity impact responses of S2-glass/SC15 epoxy composite panels with a gas gun. Int J Impact Eng 111:244–254. https://doi.org/10.1016/j.ijimpeng.2017.10.002

Vinson JR, Walker JM (1997) Ballistic impact of thin-walled composite structures. AIAA J 35:875–878. https://doi.org/10.2514/2.7461

Wang H, Hazell PJ, Shankar K et al (2017) Impact behaviour of Dyneema® fabric-reinforced composites with different resin matrices. Polym Test 61:17–26. https://doi.org/10.1016/j.polymertesting.2017.04.026

Zaera R (2011) Ballistic impacts on polymer matrix composites, composite armor, personal armor. Impact Eng Compos Struct 310–321. https://doi.org/10.1007/978-3-7091-0523-8_7

Zakikhani P, Zahari R, Sultan MTH, Majid DL (2016) Thermal degradation of four bamboo species. BioResources 11:414–425. https://doi.org/10.15376/biores.11.1.414-425

Drop Test Impact Analysis—Experimental and Numerical Evaluations

S. I. B. Syed Abdullah

Abstract The susceptibility of composites to impact damage is one of the main drawbacks for the material to be used as a structural component. This is especially true for Low Velocity Impact (LVI) loads, in which the laminate may appear pristine on surface, but considerable damage is present internally. This is often known as the Barely Visible Imapct Damage (BVID), in which this type of damage is a hidden menace which is often detrimental to the overall integrity of the structure. Therefore, it is important to fully understand the mechanics of failure to improve the damage tolerance and impact performance of the laminate. In addition, a comprehensive understand of the mechanics of failure would enable a physically based constitutive equations to be derived, thus reducing the number of experimental tests required, leading to a reduction in the total design cost. In this chapter, a review will be made on the LVI testing on composite materials, particularly on the experimental and numerical evaluations of failure and damage under LVI loadings.

Keywords Low Velocity Impact · Finite Element Methods · Damage mechanics · Fracture mechanics · Delamination

1 Introduction

The use of Fibre Reinforced Polymers (FRPs) is rapidly increasing due to its high strength-to-weight ratio. Furthermore, it's superior corrosion resistance as well as improved fatigue performance often makes it highly desirable in many industrial applications (Cantwell and Morton 1991). However, the susceptibility of FRPs to impact damage is one of the major downfall for material selection by structural designers. Low Velocity Impact (LVI), usually from large mass impactors with velocities of up to 70 m/s (Cantwell and Morton 1991; Davies and Olsson 2004), could present a significant threat to the structure. This may come in the form of in-plane fibre damage, delamination, or debonding between components, and is usually called Barely Visible Impact Damage (BVID). BVID normally cannot be detected by the

S. I. B. Syed Abdullah (✉)
Department of Aeronautics, Imperial College London, London, UK

© Springer Nature Singapore Pte Ltd. 2021
M. T. H. Sultan et al. (eds.), *Impact Studies of Composite Materials*, Composites Science and Technology, https://doi.org/10.1007/978-981-16-1323-4_3

naked eye and requires the use of Non-Destructive Testing (NDT) such as Ultrasonic C-Scan to capture the extent of damage (normally delamination).

Information obtained from experimental results are often translated into mathematical models which can then be implemented as numerical models for use in design tools such as the Finite Element Method (FEM). In recent years, the advancement in numerical modelling allows for a more accurate damage prediction, thus allowing engineers to not 'over-design' a structure to maintain its integrity under various loading conditions.

In the following sub-sections, a review will made on the impact response of composite materials under LVI loading, with an emphasis on the experimental and numerical correlation.

2   Low Velocity Impact on Composite Structures

The response of composite laminates from transverse impact loading is known to vary with the speed of impact (Abrate 1998; Davies and Olsson 2004). In Low Velocity Impact (LVI) conditions, boundary effects usually dominate since the impact duration is longer between the laminate and the impactor. Conversely, stress wave effects usually dominate in High Velocity Impact (HVI), since the impact times are usually shorter than LVI. Furthermore, damage patterns observed in the two cases are often different. In LVI, global damage modes may be observed, since large deflections often occur, which depend highly on the shear properties (both in-plane and interlaminar) of the material. Compared to LVI, the type of damage found on HVI laminates are often highly localised at the region of impact (Fig. 1).

Barely Visible Impact Damage (BVID), which occurs under LVI conditions, is often the hidden menace, causing a significant degree of damage in the composite laminate. BVID almost always result in delamination, severely compromising the integrity of the structure. (Davies et al. 1994) proposed a simple fracture mechanics-based model to predict the critical load in which delamination onset will occur in low velocity impact loading. The critical load, also called the Delamination Threshold Load (DTL), P_c, is given by:

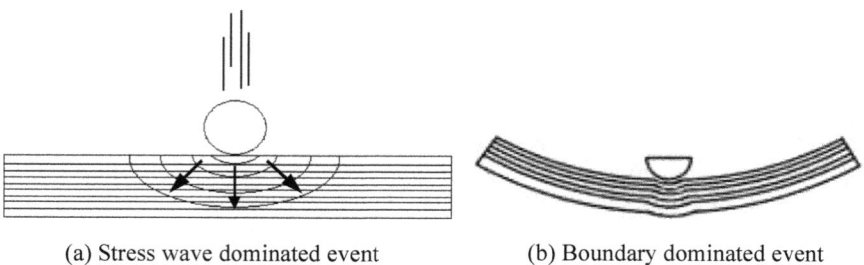

(a) Stress wave dominated event   (b) Boundary dominated event

Fig. 1 Response types for impacted composite structures (Cantwell and Morton 1991)

$$P_c^2 = \frac{8\pi^2 E_f h^3}{9(1 - v^2)} G_{IIc} \tag{1}$$

where P_c is the critical load to initiate delamination, G_{IIc} is the Mode II critical strain energy, v is the Poisson's ratio, E_f is the laminate flexural modulus, and h is the laminate thickness. The high dependence on the shear properties for LVI loading is apparent, resulting in the inclusion of Mode II critical strain energy in Eq. (1). This is not surprising since LVI is a flexural dominated event, hence interlaminar shear properties are central to the impact performance of laminated composites. However, the analytical relation proposed by Davies is only limited to classical composite types, such as Carbon Fibre/Epoxy (CF/Epoxy) and Glass Fibre/Epoxy (GF/Epoxy). In addition, Eq. (1) is also restricted to UD fabric architecture. For advanced fabric architecture, such as woven or Non-Crimp Fabrics (NCFs), Eq. (1) will yield inaccurate results.

Figure 2 (Syed Abdullah 2019) presents the LVI response for three different composite laminates, namely IM7/8552 (CF/Epoxy), S2-Glass/MTM57 (GF/Epoxy), and Vectran/MTM57 (Thermotropic Liquid Crystal Polymer, TLCP/Epoxy) under 40 J of impact energy. Note that S2-Glass/Epoxy and Vectran/Epoxy are based on the NCF architecture, whilst IM7/8552 is based on

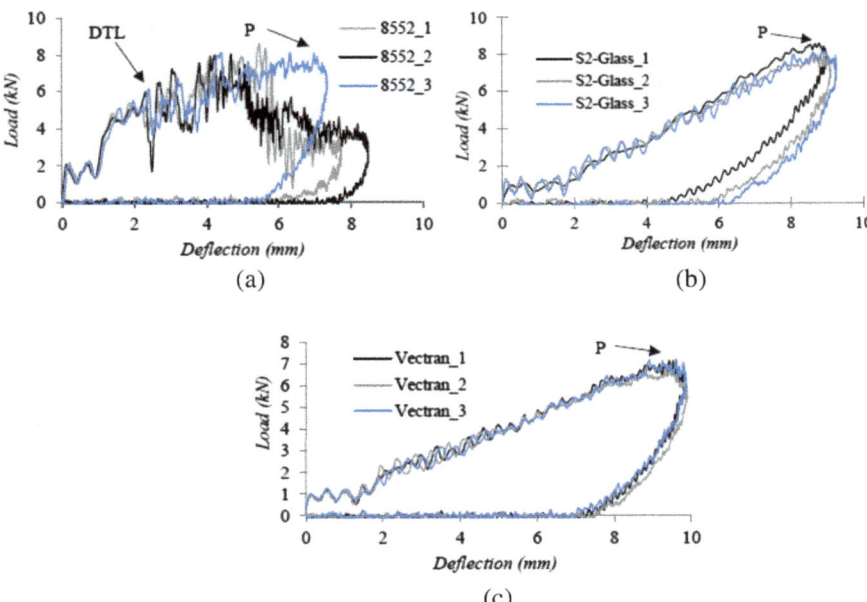

Fig. 2 LVI response for three different composite laminates under 40 J of impact energy **a** IM7/8552, **b** S2-class/MTM57, **c** vectran/MTM57 (Syed Abdullah 2019; Syed Abdullah et al. 2021)

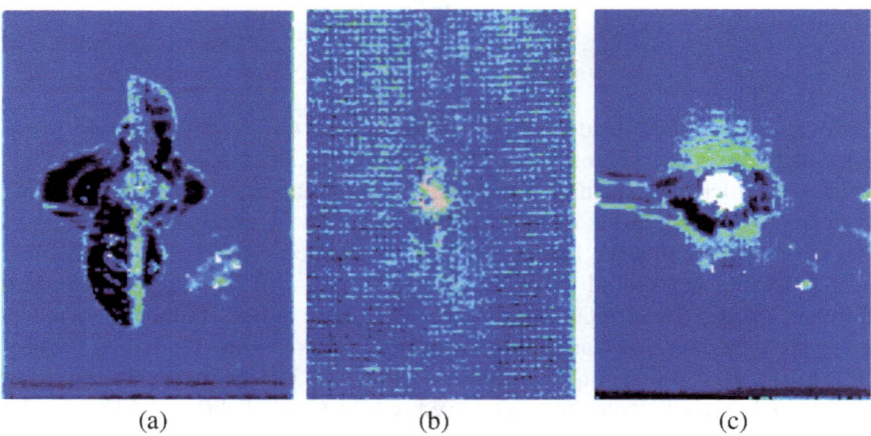

          (a)   (b)   (c)

Fig. 3 Ultrasonic C-scan image of damaged composite under 40 J LVI loading. **a** IM7/8552, **b** S2-glass/MTM57, **c** vectran/MTM57 (Syed Abdullah 2019; Syed Abdullah et al. 2021)

the UD architecture. A sharp load drop can be clearly seen in the early stage of impact in the load–deflection response of IM7/8552, Fig. 2a, whilst no observable drop can be found for S2-Glass/MTM57 and Vectran/MTM57, Fig. 2b, c, respectively. This is consistent with the findings obtained from the ultrasonic C-scan images taken from the damaged laminates, Fig. 3. From Fig. 3, it can be clearly seen that considerable delamination was present on the IM7/8552 laminate, Fig. 3a, whilst minimal delamination was observed on the S2-Glass/MTM57 and Vectran/MTM57 laminates—Fig. 3a, b, respectively.

Table 1 presents the calculated DTL using Eq. (1). Additionally, the experimentally measured DTL based on the load–deflection response presented in Fig. 2 is also included in Table 1. While the prediction of the DTL using Eq. (1) was considerably accurate for IM7/8552, this is not the case for Vectran/MTM57 and S2-Glass/MTM57 laminates. This is due to several reasons. First, the model depends on the flexural stiffness of the composite. Whilst the calculated flexural stiffness may be reliable for S2-Glass/MTM57, this may not be the case for Vectran/MTM57. The low compressive properties, both stiffness and strength, suggest that the flexural stiffness is driven by tensile strain. Secondly, the Poisson's ratio for both laminates was significantly lower than IM7/8552, due to the fabric architecture of S2-Glass/MTM57 and Vectran/MTM57, compared to IM7/8552.

Table 1 Delamination threshold load: measured and calculated (Syed Abdullah 2019)

Material	Measured DTL (kN)	Calculated DTL (kN)
IM7/8552	5.78	6.11
S2-glass/MTM57	N/A	N/A
Vectran/MTM57	N/A	N/A

Other analytical models, such as those based on the spring-mass systems (Davies and Olsson 2004), or indentation models derived from classical Hertzian contact models, are often used to capture the inelastic unloading commonly observed in tougher composites, such as those based on polymer fibres or thermoplastic matrices (Fig. 4).

An example of inelastic unloading is shown in Fig. 2a, b, where it can be observed that the load–deflection curve does not unload back directly to the origin, indicating that the laminates have absorbed a finite amount of energy transferred from the impactor during the impact event. For S2-Glass/Epoxy, damage is present in the form of matrix cracks and fibre kinking (compression failure), whilst for Vectran/MTM57, permanent indentation on the laminate front face is observed, with minimal yarn splitting and delamination on the front face of the laminate.

The analytical models described previously are often used to predict the LVI response of composite laminates. The main advantage of these models is the insight provided by closed form expressions, which directly shows the influence of different parameters. For instance, the influence of G_{IIc} in Eq. (1) is evident due to the flexural dominated response under LVI events. However, the analytical models are almost always limited to a simple geometry, and no or limited ability to model damage

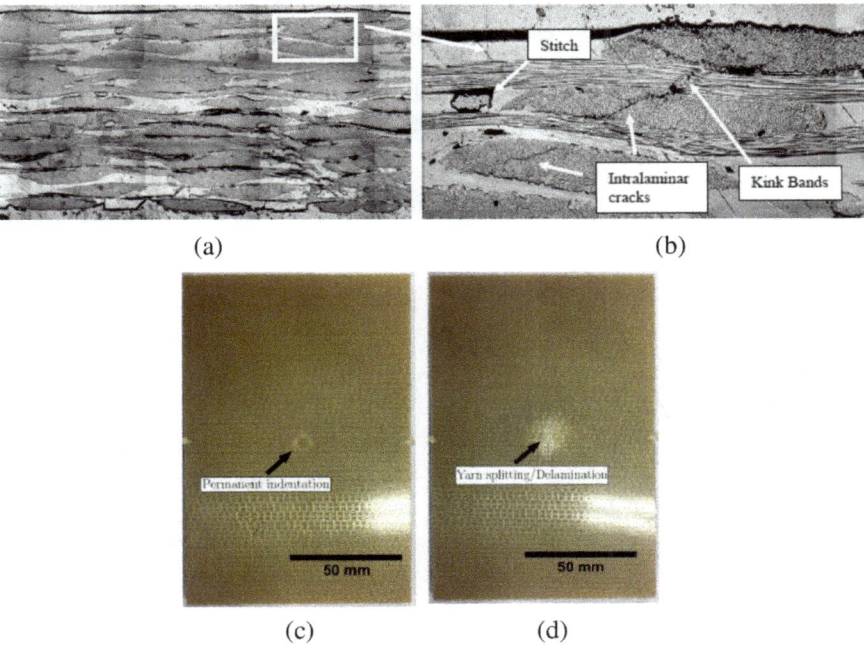

(a) (b)

(c) (d)

Fig. 4 Damage observed on impacted S2-Glass/MTM57 and vectran/MTM57 laminates, **a** cross-sectional image around the area of impact, **b** close-up image of the white square in, **a**, **c** front face of Vectran/MTM57 damaged laminate **c** back face of Vectran/MTM57 (Syed Abdullah 2019; Syed Abdullah et al. 2021)

growth. Therefore, these models are more suited to predict the impact response up to damage initiation, rather than for simulation of the actual growth. Ultimately, computational mechanics is often selected to model the full impact event, enabling a more accurate damage prediction capability.

3 Numerical Modelling of Composite Structures Under Low Velocity Impact

The use of numerical models for the prediction of mechanical properties or impact response can considerably reduce the time and cost related to composite structural design. Numerical techniques such as the Finite Element Method (FEM), is often utilised to predict damage due to impact loading in a composite material. In recent years, the use of energy-based models has seen an appreciable increase due to their accuracy and reliability in damage prediction. Mathematical models such as the Classical Laminate Theory (CLT) are used in conjuction with energy-based theories such as fracture mechanics to describe laminate failure. For example, the energy required for fibre failure is taken as the energy obtained from standard fracture mechanics tests such as the Compact Tension (CT) or Compact Compression (CC). This energy is commonly known as the strain energy release rate, G_c, and is assumed to be the area enclosed under the stress versus strain curve of the relevant modes. A typical stress–strain-damage relationship for a linear-elastic based composite laminates (such as CF/Epoxy and GF/Epoxy) is shown in Fig. 5.

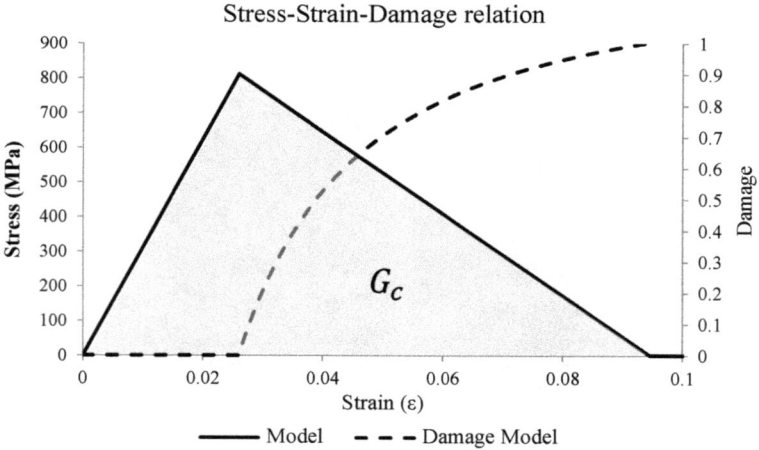

Fig. 5 Typical stress–strain-damage relationship for a linear-elastic composite laminate (Syed Abdullah et al. 2021)

The concept of degradation is essentially part of a more general Continuum Damage Mechanics (CDM) approach, which was first introduced by (Kachanov 1999) and later by (Rabotnov 1969) when attempting to describe the creep behaviour in metals. CDM is an attractive approach since it provides a method in which accurate determination of the material condition can be made—from a pristine condition (no damage) until final failure (full damage). The earlier approach of modelling laminated composites using stress (or strain) based criteria was found to be theoretically inaccurate (Tsai and wu 1971; Hashin 1980; Chang and Chang 1987), in which the stress is immediately reduced to zero upon reaching its threshold strength. This is a gross over-simplification which neglects the post-failure behaviour of a laminated composite.

The earliest implementation of energy-based damage mechanics approach is proposed by (Ladeveze and LeDantec 1992). In-plane testing of various laminate orientation was simulated using the CDM approach and later compared with the experimental results. Damage onset was taken as the failure strength of relevant modes, and then linearly degraded until zero. Excellent correlation between the experimental and simulation results was obtained. Later, (Matzenmiller et al. 1995) utilise the CDM approach by considering the post-failure behaviour as a function of the Weibull distribution of strength. (Williams and Vaziri 1995) implemented the approach suggested by (Matzenmiller et al. 1995) into LS-Dyna as a plane stress material model. Following this, Iannucci et al. (2001) (Iannucci and Ankersen 2006; Iannucci and Willows 2006, 2007; Iannucci et al. 2009) employed the CDM approach to model thin laminated composites (UD and woven) under LVI and HVI loadings. All numerical prediction including the force–time/displacement histories were in close agreement with the experimental results.

The linear degradation model, originally proposed by (Bažant and Oh 1983), has been widely used due to its ability to reproduce excellent force–time/deflection histories (Davies and Zhang 1995; Davies and Olsson 2004). However, this approach is rather simplified and does not involve any physical reasoning. For this reason, Maimi et al. (2007a, b) proposed a linear-exponential softening law, Fig. 6a, which

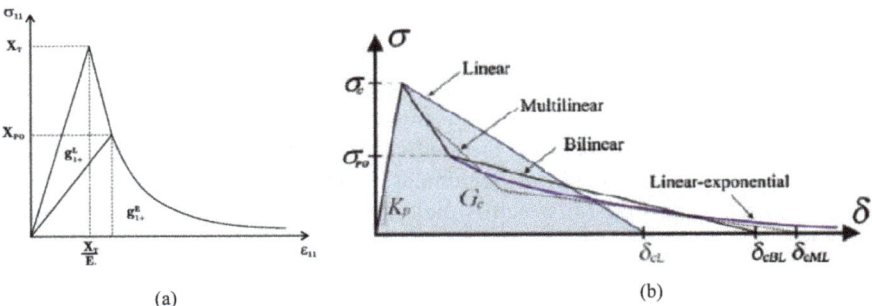

(a) (b)

Fig. 6 Alternative softening laws for fibre tension and compression failures, **a** exponential softening law, **b** linear/bi-linear/multi-linear softening laws (Dávila et al. 2009)

was later super-imposed by (Dávila et al. 2009) to a bi-linear softening law involving two linear curves to represent damage propagation—illustrated in Fig. 6b. These softening laws were based on the processes typically involved in fibre failure (i.e. matrix failure, fibre- bridging, fibre pull-out etc.). For instance, at damage onset, stress will be linearly degraded until reaching the fibre pull-out strength, X_{po}, followed by an exponential softening until zero stress.

These alternative softening laws not only provide a sound physical justification for damage degradation but were also reported to yield more accurate results. This was shown by Davila and co-workers (Dávila et al. 2009), when comparing the linear and bi-linear softening laws in a Compact Tension (CT) simulation. It was found that the bi-linear softening law was able to closely predict the shape of the Resistance curve (R-curve) with the experimental results if compared to the linear softening law.

Apart from describing an accurate determination of the material condition, the CDM approach also allows for the alleviation of mesh dependency in numerical analysis. This is because CDM is based on a representative unit volume and therefore the energy related to this unit volume must be made constant regardless of the element dimension. Thus, a length parameter is introduced, typically known as the element characteristic length, l_c, to ensure constant energy dissipation throughout the analysis. This approach is commonly known as the smeared formulation, whereby the fracture energy is smeared over the full volume of the element.

However, the calculation of l_c is not always straightforward and may require a separate algorithm in the material model. Some of the methods proposed by previous researchers include the use of the element shape functions, or purely from a geometrical approach. This was discussed by (Ehrich 2013), which investigated both strategies for the l_c calculation. For the former (using geometrical approach), the nodal coordinates from the element nodal connectivity were accessed and then stored as a history variable, which was then called back to calculate the coordinate global to local coordinate transformation based on the ply angle, θ, given by:

$$x' = \cos(\theta)x + \sin(\theta)y \tag{2}$$

$$y' = -\sin(\theta)x + \cos(\theta)y \tag{3}$$

where (x', y') are the material coordinates, and (x, y), are the global coordinates. Upon transformation, the characteristic element length, l_c can be calculated using the desired approach (geometry or element shape functions). For the geometrical approach, l_c will be calculated from the element sub-intervals giving the characteristic length for each strip, Fig. 7a, which is then used to calculate the l_c for the entire element.

Conversely, when utilising the element shape function approach, l_c, is calculated using the isoparametric nodal coordinates which is used in the partial differential equation originally proposed by (Oliver 1989) given by:

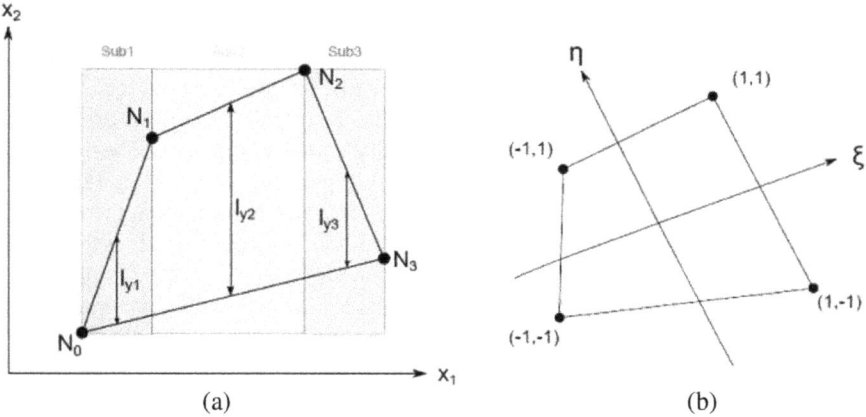

Fig. 7 Calculation of l_c from different approaches, **a** geometrical, **b** element shape function (isoparametric coordinates) (Ehrich 2013)

$$g_f = G_c \frac{\partial v(x', y')}{\partial x'} = \frac{G_c}{l_c} \tag{4}$$

where g_f is the specific fracture energy, and G_c is the critical fracture energy of the material.

Donadon et al. (2008), (Donadon and Iannucci 2006) utilised the smeared formulation to accommodate irregular/non-structured mesh, by utilising the element shape functions approach to calculate l_c for use in the smeared formulation. (Ehrich 2013) has also implemented these modifications into plane stress elements in Abaqus explicit. (Raimondo et al. 2012) proposed an alternative approach in alleviating mesh dependency. The method is based on the maximum crack density, which is suggested to be a material characteristic in a composite laminate. Hence, a crack density parameter was introduced into their degradation model to account for the total energy dissipated in one element.

4 Modelling Damage

In the context of CDM, damage was initiated when the current stress reaches the maximum strength of the material ($d = 0$), and then degraded to zero stress ($d = 1$). Each failure modes (tension, compression, and shear) are typically assigned with a damage variable, hence allowing the current state of the model to be examined. Failure normally initiate as matrix cracks and then followed by delamination, similar to the experimental observation. While the interaction between matrix failure and delamination are sometimes treated separately, a study performed by (Bouvet et al. 2012), attempts to capture this interaction by using discrete intra-ply (matrix) zero-thickness cohesive elements (or cohesive surfaces). These elements are based on non-linear springs and follow the well-known bi-linear traction–separation law, similar to the one proposed by (Bažant and Oh 1983). The authors reported a remarkable correlation between the experimental results, in spite of a couple of drawbacks. First, the approach is computationally costly due to the number of cohesive algorithms utilised in the model. Secondly, and most importantly, the authors reported a significant mesh sensitivity in the model. Perhaps it is worth noting that the model is restricted to UD materials due to the inherent nature of the approach. In a follow-up study, (Bouvet et al. 2013) utilised the previous model to simulate the Compression After Impact (CAI) response of CF/Epoxy laminates. Similar to the previous study, both the impact and the CAI response were in excellent agreement with the experimental observations. Recently, Lopes and co-workers (Lopes et al. 2016) utilised the approach proposed by (Bouvet et al. 2012) in modelling intra-ply failure. For the in-plane response, the authors employed the constitutive relations proposed by Maimi et al. (2007a, b). Although the approach is computationally costly, accurate prediction of damage types, in particular matrix cracking and delamination were achieved.

5 Conclusion

The susceptibility of composite materials to impact induced loads remains one of its greatest drawbacks. The ability of composite materials to absorb impact energies are often inferior to that of metallic materials such as steel and aluminium. One of the main reasons for this weakness is due to the large energy absorbing potential by way of plasticity of metallic materials, when compared to polymer composites. For instance, (Karthikeyan et al. 2013) investigated the ballistic performance of cured IM7/8552 (CF/Epoxy) with 304 stainless steel plates. It was concluded that the uncured 304 stainless steel plates showed significant inelastic deformation before penetration, resulting in superior ballistic performance. In contrast, the brittle nature of IM7/8552 restricts its energy absorbing potential to the elastic regime, limiting in a poor impact performance compared with the metallic system. Thus, it is essential that the performance of composite materials under impact induced loads is improved.

Since the last few decades, experimental campaigns have provided a clear insight into the mechanics of failure of composite structures under impact loads. This information enables constitutive relations to be derived, which is then used to predict failure of the composites under impact loads. In addition, the advancement of analytical and numerical techniques in recent years have greatly helped in reducing the costs associated with structural design using composite structures. The use of polymer fibre-based composites such as Vectran and Dyneema have seen a considerable increase owing to its superior performance under impact loads. However, many aspects of these composites remain unclear and not fully understood. Thus, the need to fully understand the mechanics of failure is essential for a number of reasons. First, with a comprehensive understanding, engineers could exploit the full potential of the composite, without the need to 'over-design' to ensure that the composite structure meets all safety requirements. This would result in a more cost-effective design, with minimal material consumption and fewer man-hours. Secondly, an accurate prediction of failure can be made, leading to a more reliable design since the mechanics of the composite are fully understood. Thirdly, physically based constitutive equations can be derived, which can be implemented as a material model for Finite Element Method (FEM) modelling. The material model can then be used to reduce the number of experimental tests required, leading to a reduction in the total design cost.

References

Abrate S (1998) Impact on composite structures. Impact Compos Struct. https://doi.org/10.1017/cbo9780511574504

Bažant ZP, Oh BH (1983) Crack band theory for fracture of concrete. Matériaux Constr 16:155–177. https://doi.org/10.1007/BF02486267

Bouvet C, Rivallant S, Barrau JJ (2012) Low velocity impact modelling in composite lamiantes capturing permanent indentation. Compos Sci Technol 72:1977–1988

Bouvet C, Rivallant S, Barrau JJ (2013) Failure analysis of CFRP laminates subjected to compression after impact: FE simulation using discrete interface elements. Compos Part A Appl Sci Manuf 55:83–93. https://doi.org/10.1016/j.compositesa.2013.08.003

Cantwell WJ, Morton J (1991) The impact resistance of composite materials—a review. Composites 22:347–362

Chang FK, Chang KY (1987) A progressive damage model for laminated composites containing stress concentrations. J Compos Mater 21:834–855. https://doi.org/10.1177/002199838702100904

Davies GAO, Olsson R (2004) Impact on composite structures. Aeronaut J 108:541–563. https://doi.org/10.1017/S0001924000000385

Davies GAO, Zhang X (1995) Impact damage prediction in carbon composite structures. Int J Impact Eng 16:149–170. https://doi.org/10.1016/0734-743X(94)00039-Y

Davies GAO, Zhang X, Zhou G, Watson S (1994) Numerical modelling of impact damage. Composites 25:342–350. https://doi.org/10.1016/S0010-4361(94)80004-9

Dávila CG, Rose CA, Camanho PP (2009) A procedure for superposing linear cohesive laws to represent multiple damage mechanisms in the fracture of composites. Int J Fract 158:211–223. https://doi.org/10.1007/s10704-009-9366-z

Donadon MV, Iannucci L, Falzon BG et al (2008) A progressive failure model for composite laminates subjected to low velocity impact damage. Comput Struct 86:1232–1252. https://doi.org/10.1016/j.compstruc.2007.11.004

Donadon MV, Iannucci L (2006) An objectivity algorithm for strain softening material models. In: 9th international LS-DYNA users conference, pp 43–54

Ehrich F (2013) Low velocity impact on pre-loaded composite structures

Hashin Z (1980) Failure of unidirectional fibre composites. J Appl Mech 47:329–334

Iannucci L, Dechaene R, Willows M, Degrieck J (2001) A failure model for the analysis of thin wove glass composite structures under impact loadings. Comput Struct 79:997–1011

Iannucci L, Ankersen J (2006) An energy based damage model for thin laminated composites. Compos Sci Technol 66:934–951. https://doi.org/10.1016/j.compscitech.2005.07.033

Iannucci L, Pope DJ, Dalzell M (2009) A constitutive model for dyneema UD composites. In: ICCM international conference on composite materials

Iannucci L, Willows ML (2006) An energy based damage mechanics approach to modelling impact onto woven composite materials—part I: numerical models. Comput Part A 37:2041–2056

Iannucci L, Willows ML (2007) An energy based damage mechanics approach to modelling impact onto woven composite materials: part II. Experimental and numerical results. Compos Part a Appl Sci Manuf 38:540–554. https://doi.org/10.1016/j.compositesa.2006.02.023

Kachanov LM (1999) Rupture time under creep conditions. Int J Fract 97. https://doi.org/10.1023/A:1018671022008

Karthikeyan K, Russell BP, Fleck NA et al (2013) The effect of shear strength on the ballistic response of laminated composite plates. Eur J Mech A Solids 42:35–53. https://doi.org/10.1016/j.euromechsol.2013.04.002

Ladeveze P, LeDantec E (1992) Damage modelling of the elementary ply for laminated composites. Compos Sci Technol 43:257–267. https://doi.org/10.1016/0266-3538(92)90097-M

Lopes CS, Sádaba S, González C et al (2016) Physically-sound simulation of low-velocity impact on fiber reinforced laminates. Int J Impact Eng 92:3–17. https://doi.org/10.1016/j.ijimpeng.2015.05.014

Maimí P, Camanho PP, Mayugo JA, Dávila CG (2007a) A continuum damage model for composite laminates: part I—constitutive model. Mech Mater 39:897–908. https://doi.org/10.1016/j.mechmat.2007.03.005

Maimí P, Camanho PP, Mayugo JA, Dávila CG (2007b) A continuum damage model for composite laminates: part I—constitutive model. Mech Mater 39:909–919. https://doi.org/10.1016/j.mechmat.2007.03.005

Matzenmiller A, Lubliner J, Taylor RL (1995) A constitutive model for anisotropic damage in fiber-composites. Mech Mater 20:125–152. https://doi.org/10.1016/0167-6636(94)00053-0

Oliver J (1989) A consistent characteristic length for smeared cracking models. Int J Numer Methods Eng 28:461–474. https://doi.org/10.1002/nme.1620280214

Rabotnov YN (1969) Creep rupture. In: Proceedings of the 12 international congress on applied mechanics. Stanford

Raimondo L, Iannucci L, Robinson P, Curtis PT (2012) A progressive failure model for mesh-size-independent FE analysis of composite laminates subject to low-velocity impact damage. Compos Sci Technol 72:624–632. https://doi.org/10.1016/j.compscitech.2012.01.007

Syed Abdullah SIB (2019) The impact behaviour of high performance fibre composites. Imperial College London, London

Syed Abdullah SIB, Iannucci L, Greenhalgh ES, Ahmad Z (2021) The impact performance of Vectran/Epoxy composite laminates with a novel non-crimp fabric architecture. Comp Struct 265:1–17. https://doi.org/10.1016/j.compstruct.2021.113784

Tsai SW, Wu EM (1971) A general theory of strength for anisotropic materials. J Compos Mater 5:58–80. https://doi.org/10.1177/002199837100500106

Williams KV, Vaziri R (1995) A constitutive model for anisotropic damage in fiber-composites. Mech Mater 20:125–152. https://doi.org/10.1016/0167-6636(94)00053-0

Post Impact Behavior and Compression After Impact Properties of Polymers and Their Composites—A Review

I. Siva, Avinash Shinde, I. Sankar, Chithirai Pon Selvan, and M. T. H. Sultan

Abstract Understanding the post-impact properties are essential while designing a new material system. Serviceability and sustainability are matters to protect the assembly from accidents since transportation and assembly operations experience many micro-damages in the material system. Researchers carried out several experiments to understand the post-impact performance of the laboratory scale specimens. Upon conducting the low velocity and other impact studies, the tensile, compression after impact (CAI), and related static properties have been measured and reported by the researchers. Even though the exploration is wider to analyze in the literature, a categorization is needed to proceed the further research in this field. Challenges are classified as conducting experiments, measuring the data, visualizing the results, data correlation and interpretation. Present review also contexted in a similar fashion to create a deeper insite about the post-impact characteristics.

I. Siva (✉)
Department of Mechanical Engineering, Centre for Composite Materials, Kalasalingam Academy of Research and Education, Anand Nagar, Tamil Nadu, India
e-mail: isiva@klu.ac.in

A. Shinde
Department of Mechanical Engineering, Cummins College of Engineering, Pune, Maharashtra, India

I. Sankar
Department of Mechanical Engineering, National Engineering, Kovilpatti, Tamil Nadu, India

C. P. Selvan
School of Science and Engineering, Curtin University Dubai, Dubai, United Arab Emirates

M. T. H. Sultan
Aerospace Manufacturing Research Centre (AMRC), Faculty of Engineering, Universiti Putra Malaysia (UPM), 43400 UPM Serdang, Selangor Darul Ehsan, Malaysia

Department of Aerospace Engineering, Faculty of Engineering, UPM, 43400 UPM Serdang, Selangor Darul Ehsan, Malaysia

Laboratory of Biocomposite Technology, Institute of Tropical Forestry and Forest Products (INTROP), UPM, 43400 UPM Serdang, Selangor Darul Ehsan, Malaysia

M. T. H. Sultan
e-mail: thariq@upm.edu.my

© Springer Nature Singapore Pte Ltd. 2021
M. T. H. Sultan et al. (eds.), *Impact Studies of Composite Materials*, Composites Science and Technology, https://doi.org/10.1007/978-981-16-1323-4_4

Keywords Post-impact behaviors · Compression after impact · Low velocity impact · Review

1 Introduction

As impact damages are unavoidable in different applications, so it becomes inevitable and essential to the study of after impact analysis for the safety of structures. Following literature summary shows the after-impact studies performed on different composites. Majority of authors have focused their studies on the compression after impact (CAI) for different configuration of materials. The structural load bearing capacity is severaly reduced due to the micro-cracks induced during the low velocity impacts, resulting in catrascopic failure. Hence it is most essential to understand the post impact behaviors before installing a new material into a structure. Researchers have explored the change in materials behaviour after impact as a function of material type, loading nature, time etc.

Fiber reinforced composites are sensitive to impact and thereby reduces the mechanical properties especially compression performance after impact. To study this author performed the compression after impact (CAI) test on the fiber reinforced composites. Local buckling and shear failures are observed in compression test. As impact energy increases the compressive strength and failure strain decreases due to impact damage. Due to impact the damage area increases resulting in reduced bearing area and in turn reduction in bearing capacity. The impact damage sharply reduces compressive strength and stiffness of the material, and the residual compression strength and stiffness were decreased to 62.25 and 76.6%, respectively (Hongji et al. 2019).

1.1 Constituent Type as Factors

Several reports are in literature to characterize the significance of constituent types especially the type of reinforcement on the post-impact behaviours. The load distribution mechanism is one of the key factor to study when the post-impact in concern. An Investigation of unstitched and stitched ([02/902] S-type A and [902/02] S-type B) laminated samples is done by applying low velocity impact energies 1–8 J. After impact flexural tests using three-point bending test are carried out. The flexural properties are observed to be greatly affected due to impact. Due to stitching in direction parallel to fiber direction some damage may be induced which revealed by reduction in flexural strength for both stitched and unstitched specimens (Francesconi and Aymerich 2018). A conclusion arrived as to sustain higher post-impact strength, the secondary stitching can be introduced as normal to the fiber impregmated direction.

Low velocity impact response of 3D integrated-woven hybrid sandwich composite panels is studied in this paper. Damage in terms of compressive strength loss due to impact is interpreted. The hybrid sandwich showed lower impact index but residual strength is observed to be higher than the glass/epoxy composites. Two different core thicknesses and three different pile thickness are used for preparation of samples. Author reported that as piles thickness reduces the maximum impact force whereas as pile density increases peak load also increases.

Impact and compression properties are studied for woven composites after thermal aging degradation at 180 degree and for different days. The compression after impact properties are greatly reduced after thermal aging at 32 days. Interface damage and matrix degradation are the important factor responsible for the degradation in properties. Damage evolution study by FEA revealed brittle and progressive failure modes for unaged and aged samples respectively and author expects that, this might be due to stress plateau in stress–strain curve (Cao et al. 2018). Un treated woven flax fiber reinforced PLA composites were prepared through compression moulding technique. Low velocity impact test was conducted on the composite samples followed by the after impact behavior evaluation. The test results were compared with the carbon/epoxy laminates. The comparison revealed the advantages of flax/PLA composites over carbon/epoxy laminates in view of the energy absorption and normalized residual strength. The flax/PLA composites possessed higher energy absorption than the carbon/epoxy laminates due to the absence of delamination. The only failure mode noted in the flax/PLA composites was fiber breakage which was having no relation with the energy absorption and residual strength. Besides, carbon/epoxy laminated suffered due to the delamination and thus the residual strength was pulled down.

1.2 Role of Hybridization

Another study, an external hybrid patches of glass/Kevlar are used to repair the damaged composite on low velocity impact and the quasi-static tensile test response is studied after impact response. Different impact energies are used for the study and it is observed that hybridization influences the energy absorption. Intra-ply hybrid patches showed better impact properties (Andrew et al. 2019). In this, author developed a novel sandwich panel with hybrid skin layers and core made of foam. The skin is hybridization of woven glass and wire net in an epoxy matrix. Compression after impact tests is performed to determine residual properties after impact. Author reported great enhancement in impact and residual properties. In impact, three damage modes viz. breaking, delamination and foam cracking, while in compression after test two damage modes viz. buckling and delamination are observed. Digital image correlation and SEM are used for damage mechanism analysis (Wan et al. 2020).

The pile hybridization reduces the core stiffness which resulted in reduction in maximum impact force (Mirdehghan et al. 2020). E-glass/S-glass/Epoxy hybrid sandwich with PVC as core material is fabricated. Different stacking sequence and impact velocities are used for this study. Post impact flexural strength is measured.

The author observed that the carbon fibers in the face sheet increased the peak load and impact energies. It is also observed to increase the flexural strength of composite (Özen 2017). Author has investigated the compression after low velocity impact (CAI) properties of stitched and unstitched composite. Samples with hand stitching and machine stitching are prepared. According to ASTM D7136 test were conducted and ASTM D7137 is used for compression after low velocity impact test. Optical microscope and digital image analyzer software is used for damage analysis. Use of mechanical needle caused filament breakage and that affected the CAI and damage sizes. Author observed that the stitched specimens have higher impact strength than unstitched specimens although multi-stitched damage areas are larger. Author suggested this is because multi-stitching prevents inter-layer delamination in out of plane direction and intra-splitting of fiber in in-plane direction (Erdogan and Bilisik 2018).

Impact and after impact behavior of carbon/epoxy laminate modified with milled glass fiber are investigated in this article. Two different layup configurations viz. unidirectional (UD) and cross-ply (CP) are evaluated for different impact velocities. The residual load bearing capacity of post impacted specimens is evaluated by three point bending test. The filled samples showed higher peak force, lower deformation and lower damage the milled glass fiber prevents the crack propagation and reduces the damage. The residual flexural after impact (FAI) is obvious to decrease with increasing velocities. But filled samples show less reduction in FAI as compared to unfilled samples. Author also observed that absorption of impact energy is more in CP samples. CP filed and UD filled specimens are better for low impact energies and high impact energies respectively (Kannivel et al. 2019).

A low velocity impact property of Kevlar/basalt polypropylene composite is investigated. Two different compositions with different staking sequence are prepared. Drop weight impact tests are carried out with 25, 50 and 75 J energies. Hybrid with alternate layers of Kevlar and basalt exhibited higher impact energy absorption. These experimental results are compared with the numerical simulation done in Abaqus and found in agreement (Bandaru et al. 2018).

Low velocity impact properties of pure epoxy and epoxy/glass composites are studied this paper. The composite is reinforced with MWCNT with different weight percentages dispersed in epoxy matrix by sonication process. Considerable improvement in the impact properties is observed with addition of small percentage of MWCNT. 0.34% CNT showed highest improvement (Ranjbar and Feli 2019).

The study of microstructure and residual strength of glass fiber reinforced composite with different weaving pattern after impact is done in this paper. More plies are damaged in cross ply composite than plain-weave composite so plain-weave showing better impact properties than the cross-ply composite. However plain-weave exhibited less residual strength under high impact energy. As a conclusion author reported matrix cracking and delamination happened due to low velocity impact (Liu et al. 2020).

2   Fillers and Nanoparticles Incorporations

Fillers especially in nano-scale restrict the molecular movements in the polymeric system during strain. Most of these tiny particles have less absorption characteristics whereas the carbon nanotube kind holo structures may posses higher absorption behavior. Hence the incorporation of the these particles into the polymers or polymer composite may impart additional post impact stability.

A new material epoxy/glass Nano composite with 10 wt% Nano silica and 2.5 wt% Nafen alumina Nano fiber are fabricated using VARTM. Impact strength, energy absorbed and damage area for the two Nano fillers is compared at different impacting energies. Nano silica and alumina Nano fiber showed rigid and stiff behavior respectively, at higher impacting energy. The author reported higher pick force in Nano filled composites compared to non-filled (Kallagunta and Tate 2019).

Cast iron and with different percentage of bronze chips are used to fabricate the metal matrix composite using hot isostatic pressing. A drop weight test with loading rate at 2 m/s was performed and results are compared with bulk iron and bronze, individually. The author observed that MMCs can be used to indicate bulk material properties even with porosity of 2–8% (Şahin et al. 2019). Functionally graded carbon nanotubes laminate with different stacking sequence is studied to evaluate the low velocity impact performance of the composite. Higher impact velocity is observed to give larger deformation (Yang et al. 2018).

In contrast, Nor et al. (2019) have built natural fiber composite with dopped carbon nanotubes in various weight portions. Authors have conducted low velocity impact studies on the plain and nanohybrid composites to characterize the effect of nanofiller incorporation on the LVI and after impact properties. In LVI test, the composites incorporated with the CNT consumed less energy and produced smaller cracks compared to the another manufactured one with no nanofillers. To characterize the impact damage, the ultrasonic wave propagation imaging technique is employed. Authors have reported that a 23% increment in the after impact strength were noted with the samples prepared with nanofillers. Especially the nanocomposites produced at 10% filler addition have shown superior sustainabilities among the produced composites. Further the gain is linearly decreased with the increase in impact load is also reported in the research article.

Unlike the reviewed articles, Gliszczynski et al. (2020) experimented the CAL of the channels as similar to the previous study by applying larger impact loads. Studies were conducted at 20 and 30 J load for low velocity impact on the polymer composites. Laminated samples were groups as quasi-isotropic, quasi-orthotropic and angle ply arrangements and experiments conducted. According to the authors results the most unsafe impact would be the damage caused at corner which is perpendicular to the web. The compression after impact properties of the corner damaged composites are found weaker. Numerical modeling on the CAI also supports the experimental conclusion in a great deal. An unsymmetrical load buckling was caused by the corner impact in all the 20 and 30 J tested samples and a prebuckling studies were proposed in the conclusion.

3   Sandwich Composites

Compose of spongy foams or layered honeycombs in core and cladded with the polymeric layers or polymer composite sheets as skin are the sandwich structures. The superior shock absorption capacity of the core material uplift the impact strength of the polymer composite. A similar response is expected in the post-impact characteristic of the sandwich composites and results of the several researchers have support the anticipations.

In their research, author used aluminum honeycomb cells and glass and carbon fiber reinforced composited for face sheet. Different thicknesses of core and face sheets are compared for low velocity impact behavior. Author noted some important observations, as the face thickness increases the impact strength of the composite increased. However, increasing the core thickness did not increase the energy absorption but damage depth is found to be increased (Topkaya and Solmaz 2018).

Low velocity impact study of hybridized carbon/basalt fiber reinforced composite. Neat carbon and basalt composite with 60% volume fraction and 60% wt fraction of carbon and basalt in equal part are prepared. Drop test at different energies is carried out and effect of hybridization on contact force and energy absorbed are investigated. SEM images showed different damage modes as micro cracks, delamination, fiber pull out and fiber breakage. Author reported the improvement in the impact properties with hybridization. However the dominant modes in CFRP and BFRP are fiber breakage and matrix damage respectively (Shishevan and Akbulut 2019).

Low velocity impact tests are performed on carbon fiber reinforced poly-methylmethacrylate (PMMA) modified with Nano fillers. Ultrasonic NDT is done to find out after impact delamination. Nano-modified composites are observed to absorb 10% more energy than unmodified composite. Author justifies that is because of more interaction of Nano fillers with PMMA (Žukienė et al. 2019).

Carbon fiber reinforced composites (CFRP) with two different resins vynilester and epoxy is investigated in this article for low velocity impact tests. Different impacting energies are used for evaluation of penetration energy and damage propagation. After impact non-destructive tests are carried out to measure the damage. Vynilester based composite shows better results as compared to epoxy based composite (Papa et al. 2019).

Wu and Wan (2019) have developed hybrid sandwich structures using glass fiber reinforced epoxy face sheets and foam core. The face sheets were embedded with and without shape memory (SMA) alloy/conventional 304SS wire nets and the specimens were designed with different orientations and numbers of SMA wires. Vacuum assisted resin infusion technique was followed to prepare the sandwich panels. The sandwich structures were undergone to low velocity impact tests and the results were reported. Digital image correlation technology was used to conduct the After-impact compression tests. The failure mechanisms were analyzed through visual inspections and scanning electron microscopy. The test results revealed that, the sandwich panels embedded with SMA wire were able to absorb more impact energy and possessed better compression after impact characters due to the super elastic behavior of the

SMA wire. More impact energy dissipation was observed in the sandwich panels embedded with SMA wires of cross configuration. The dissipated kinetic energy was transferred to the outer region of the upper face-sheet and the impact resistance of the panels was increased. The failure during the after impact compression test was initiated by the delamination between the face sheet and foam core in the impact region. The propagation of the delamination further reduced the compression carrying capacity of the structure and maximum strain can be identified at the damaged region of the face-sheet.

4 Conclusion

This review article highlighting the importance of evaluating the after-impact compression behavior on engineering materials. Many times, the low velocity impact severely degrades the structural properties of the composites. In particular, the presence of micro damages caused by low velocity impact will seriously reduce the compressive strength of the composite materials. Hence, the impact damage tolerance of primary composite aircraft structures observes more attention in the design of airframe. To evaluate the after-impact compression behaviors, two major issues are need to be considered. First, due to the complexity of the failure mechanism of impact damages, the damage must be simplified on the basis of compressive failure mechanisms of the composites. Second, the failure evaluation of impact damage of composites must be standardized. By these systematic approaches, the after-impact studies are can be conducted to evaluate the compressive load bearing capacity of the materials.

Review concludes that, the research on the CAI and related can be carried by accounting the fiber type, pattern, stacking sequence, aging and filler incorporation. On the other hand, harnessing the modeling tool to assess the compression after impact and other post-impact properties will save the time and materials.

References

Andrew JJ, Srinivasan SM, Arockiarajan A (2019) Influence of patch lay-up configuration and hybridization on low velocity impact and post-impact tensile response of repaired glass fiber reinforced plastic composites. J Compos Mater 53:3–17. https://doi.org/10.1177/002199831877 9430

Bandaru AK, Patel S, Ahmad S, Bhatnagar N (2018) An experimental and numerical investigation on the low velocity impact response of thermoplastic hybrid composites. J Compos Mater 52:877–889. https://doi.org/10.1177/0021998317714043

Cao M, Wang H, Gu B, Sun B (2018) Impact damage and compression behaviours of three-dimensional angle-interlock woven composites after thermo-oxidation degradation. J Compos Mater 52:2085–2101. https://doi.org/10.1177/0021998317740199

Erdogan G, Bilisik K (2018) Compression after low-velocity impact (CAI) properties of multi-stitched composites. Mech Adv Mater Struct 25:623–636. https://doi.org/10.1080/15376494.2017.1308601

Francesconi L, Aymerich F (2018) Effect of stitching on the flexure after impact behavior of thin laminated composites. Proc Inst Mech Eng Part C J Mech Eng Sci 232:1374–1388. https://doi.org/10.1177/0954406217712745

Gliszczynski A, Bogenfeld R, Degenhardt R, Kubiak T (2020) Corner impact and compression after impact (CAI) of thin-walled composite profile—an experimental study. Compos Struct 248:112502. https://doi.org/10.1016/j.compstruct.2020.112502

Hongji Z, Yuanyuan G, Hong T et al (2019) Examination of low-velocity impact and mechanical properties after impact of fiber-reinforced prepreg composites. Polym Compos 40:2154–2164. https://doi.org/10.1002/pc.25002

Kallagunta H, Tate JS (2019) Low-velocity impact behavior of glass fiber epoxy composites modified with nanoceramic particles. J Compos Mater. https://doi.org/10.1177/0021998319893435

Kannivel S, Periasamy K, Vellayaraj A (2019) Impact response and post-impact performance of unidirectional and crossply carbon/epoxy laminates modified with milled glass fibers. Polym Compos 40:2441–2451. https://doi.org/10.1002/pc.25114

Liu W, Zhang H, Feng H et al (2020) Effect of fiber architecture on the residual strength of laminate glass fiber-reinforced polymer composites after impact. Adv Compos Lett 29:2633366X1989791. https://doi.org/10.1177/2633366x19897919

Mirdehghan A, Nosraty H, Shokrieh MM, Akhbari M (2020) Manufacturing and drop-weight impact properties of three-dimensional integrated-woven sandwich composite panels with hybrid core. J Ind Text 1–29. https://doi.org/10.1177/1528083719896764

Nor AFM, Sultan MTH, Jawaid M et al (2019) Analysing impact properties of CNT filled bamboo/glass hybrid nanocomposites through drop-weight impact testing, UWPI and compression-after-impact behaviour. Compos Part B Eng 168:166–174. https://doi.org/10.1016/j.compositesb.2018.12.061

Özen M (2017) Influence of stacking sequence on the impact and postimpact bending behavior of hybrid sandwich composites. Mech Compos Mater 52:759–766. https://doi.org/10.1007/s11029-017-9626-3

Papa I, Langella A, Lopresto V (2019) CFRP laminates under low-velocity impact conditions: influence of matrix and temperature. Polym Eng Sci 59:2429–2437. https://doi.org/10.1002/pen.25102

Ranjbar M, Feli S (2019) Mechanical and low-velocity impact properties of epoxy-composite beams reinforced by MWCNTs. J Compos Mater 53:693–705. https://doi.org/10.1177/0021998318790049

Şahin ÖS, Güneş A, Aslan A et al (2019) Low-velocity impact behavior of porous metal matrix composites produced by recycling of bronze and iron chips. Iran J Sci Technol Trans Mech Eng 43:53–60. https://doi.org/10.1007/s40997-017-0139-4

Shishevan FA, Akbulut H (2019) Low-velocity impact behavior of carbon/basalt fiber-reinforced intra-ply hybrid composites. Iran J Sci Technol Trans Mech Eng 43:225–234. https://doi.org/10.1007/s40997-018-0151-3

Topkaya T, Solmaz MY (2018) Investigation of low velocity impact behaviors of honeycomb sandwich composites. J Mech Sci Technol 32:3161–3167. https://doi.org/10.1007/s12206-018-0619-5

Wan Y, Yang B, Jin P et al (2020) Impact and compression-after-impact behavior of sandwich panel comprised of foam core and wire nets/glass fiber reinforced epoxy hybrid facesheets. J Sandw Struct Mater 1–24. https://doi.org/10.1177/1099636220912785

Wu Y, Wan Y (2019) The low-velocity impact and compression after impact (CAI) behavior of foam core sandwich panels with shape memory alloy hybrid face-sheets. Sci Eng Compos Mater 26:517–530. https://doi.org/10.1515/secm-2019-0034

Yang CH, Ma WN, Ma DW (2018) Low-velocity impact analysis of carbon nanotube reinforced composite laminates. J Mater Sci 53:637–656. https://doi.org/10.1007/s10853-017-1538-z

Žukienė K, Žukauskas E, Kažys RJ et al (2019) Structure—impact properties relationships of carbon fiber reinforced poly(methyl methacrylate) composite. Polym Compos 40:E333–E341. https://doi.org/10.1002/pc.24665

Damage and Failure in Composite Structures

S. I. B. Syed Abdullah

Abstract Failure of a composite structure under impact loads is often very complex due to the presence of many concurrent phenomena occurring during the event. For some fibres, such as polymer fibre-based composites, the types of damage may be difficult to identify characterise. In general, damage under impact loads can be identified into three different types—matrix failure, delamination, and fibre failure. Often, the use of Non-Destructive Techniques (NDTs) such as the ultrasonic C-scan and X-ray radiography is useful to identify damage inside the laminate, especially those due to Barely Visible Impact Damage (BVID). The progression of damage may also be gleaned using fractographic techniques to understand the micro-mechanics of failure in the laminate. In this chapter, failure of composite structures would be reviewed, beginning with in-plane failures (tension and compression), delamination damage, and finally damage due to impact. A discussion on the use of NDTs to identify damage, as well as fractographic techniques to understand the progression of damage will also be made.

Keywords Impact Damage · Fractography · Delamination · Fibre failure · Interlaminar fracture toughness

1 Failure in Composite Materials—An Introduction

While the first use of composite structures dates back to 1500 B.C., their significant development began in 1940 (Roseler et al. 2007), when Glass Fibre Reinforced Plastics (GFRPs) were used in the marine industry to replace traditional wood or metal structures. The appealing properties such as a high stiffness-to-weight ratio, corrosion resistant, and increased fatigue life have attracted many designers to adopt GFRP in the production of boats and ships. In aeronautical structures such as commercial aircraft, the use of composite materials has seen a rapid increase beginning in the late 1980s. Airbus introduced large primary aerodynamic structures such as the vertical and horizontal stabiliser built from Carbon Fibre Reinforced Plastics (CFRPs) in

S. I. B. Syed Abdullah (✉)
Department of Aeronautics, Imperial College London, London, UK

© Springer Nature Singapore Pte Ltd. 2021
M. T. H. Sultan et al. (eds.), *Impact Studies of Composite Materials*, Composites Science and Technology, https://doi.org/10.1007/978-981-16-1323-4_5

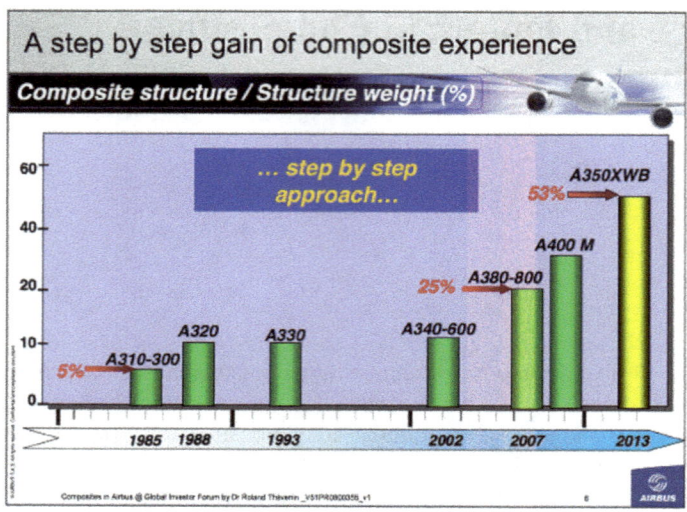

Fig. 1 Percentage use of composite materials for the past 30 years (Hellard 2008)

its A320 family. Today, the Airbus A350 XWB and its American counterpart, the Boeing 787 Dreamliner, are constructed of more than 50% composite materials (Hellard 2008). Figure 1 shows the evolution of composite material used in Airbus commercial aircraft families over the past 30 years.

The rapid increase in the use of composite materials within structural components has also stimulated a significant amount of research. Focus has been on understanding of the mechanics of the composite, as well as the ability to predict failure under a range of loading conditions. This includes impact, which can generate a considerable amount of damage in the composite. In the aerospace industry, operational threats during service, such as runway debris (Nguyen et al. 2008, 2014), bird strike (Dennis and Lyle 2009; Hedayati and Sadighi 2015), and hail ice impact (Juntikka and Olsson 2009; Kim et al. 2003) have the potential to cause a severe damage on the primary aircraft structure, incurring very expensive costs, and in extreme cases, severe injuries or even fatalities. Therefore, a comprehensive understanding of the composite material behaviour is important towards a safer and a more cost-effective aircraft design.

Failure of composite structures is often a complex process and require a significant amount of time to examine and understand. This is particularly true for high performance fibres, where the mechanics of failure have yet to be thoroughly understood. For instance, the use of polymer-based fibres such as Vectran and Dyneema in the defence and aerospace industries have seen a considerable increase in the past few decades. Owing to this, research on polymer-based fibre composites have also undergone an appreciable increase. However, many aspects of these composites remain unclear and not fully understood. For example, although the impact performance of polymer fibre-based composites has been well documented, their delamination behaviour has yet to be studied. This is an important area since delamination resistance of a composite material often determines its impact performance.

2 Fibre and Matrix Dominated Failure

Often, catastrophic failure of a composite will involve some type of fibre fracture mode, or a combination of fibre and matrix damage. This is because the fibres in a composite is the main load bearer, thus, are expected to withstand loads which are applied to the laminate. The role of the matrix in a composite is equally as important due to several reasons. Firstly, the choice of the matrix would determine the performance of the composite under certain situations. For instance, under impact loadings, thermoplastic matrices are more favourable, due to it being much tougher compared to thermosetting matrix, hence enabling to absorb a higher impact energy (Davies and Olsson 2004; Greenhalgh2009; Greenhalgh et al. 2013). Secondly, the use of thermoplastic matrices in a composite would result in a significant improvement in the interlaminar fracture toughness, hence delamination damage will be minimised (Greenhalgh 2009; Wisnom 2012). Thirdly, the use of thermoplastic matrices in a composite is not suitable for high temperature applications due to its low melting temperature, T_m, compared to thermosetting matrices (Greenhalgh 2009; Wang et al. 2018). In addition, due the low melting temperature of thermoplastic matrices, including thermoplastic fibres such as Dyneema and Vectran, frictional effects may cause the local fibre temperature to rise hence resulting in a slight degradation of the fibre properties. This effect was observed by Greenhalgh et al. (2013), when investigating the impact performance of Dyneema HB26 under High Velocity Impact (HVI) loadings. Lastly, it was found that the interfacial properties between the fibre and matrix for thermosetting matrices was much superior compared to that of thermoplastic matrices (Greenhalgh 2009; Huang and Young 1996). This is explained by the lack of chemical bonding in thermoplastic matrices, if compared its thermosetting counterpart.

In general, there are five main failure types in a composite, namely, in-plane failure, out-of-plane failure, interlaminar, intralaminar, and translaminar failure. The three latter failure types are often related to fracture-based failure. For instance, interlaminar failure is normally associated with failures in-between each composite ply, whereas intralaminar failure is often related to failure in the through-thickness direction of a laminate. Finally, translaminar failure is normally associated to failures involving fibre (tensile and compressive). An illustration of the interlaminar, intralaminar, and translaminar failures are provided in Fig. 2.

Alternatively, in-plane failures include tension, compression, and in-plane shear, whilst out-of-plane failures are related to flexure or impact-based failures. The next sub-section will provide an overview of the in-plane failure (tension and compression), and fracture-based failures. Focus will be given on interlaminar and impact damage due to the importance of these types of damage in impact loading.

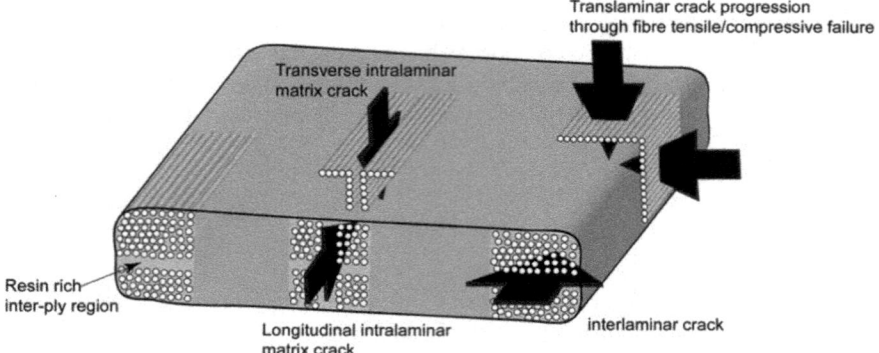

Fig. 2 Overview of ply-level failure modes (Laffan et al. 2012)

2.1 Tensile Damage

Perhaps the simplest failure mode when examining laminated composites is tensile failure. A large number of research has been performed to understand the mechanisms associated with this mode (Greenhalgh 2009; Zhuang et al. 2016; Rosen 1964; Zweben 1968; Okabe et al. 2001).

At a microscopic level, failure would initiate at an internal defect, emanating radially towards the edge of the fibre, shown in Fig. 3a. Damage would then propagate through the matrix, before reaching the adjacent fibres. This process is commonly known as the Cook-Gordon mechanism (Cook and Gordon 1964; Hull and Clyne 1996), illustrated in Fig. 3b. Consider a tension crack propagating through the matrix. Upon reaching a fibre, the shear stress at the tip of the crack will be parallel to the fibre, resulting in a shear stress at the fibre/matrix interface. If the bonding strength between the fibre and the matrix is high, the degree of fibre debonding will be limited, hence leading to a relatively planar fibre surface. In contrast, if the bonding strength

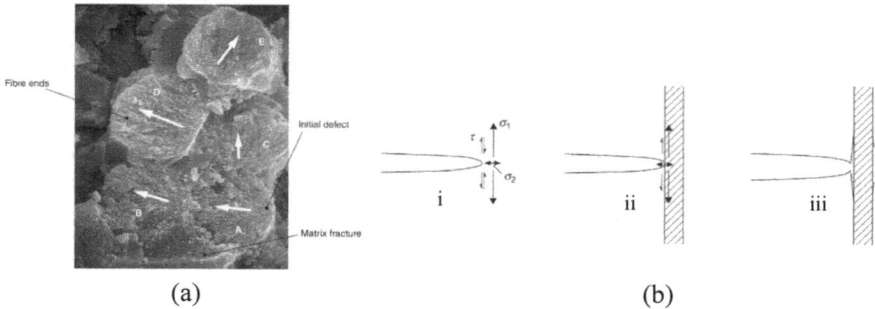

(a)   (b)

Fig. 3 **a** Failure morphology of CF/epoxy under tensile loads **b** the Cook-Gordon mechanism (Greenhalgh 2009)

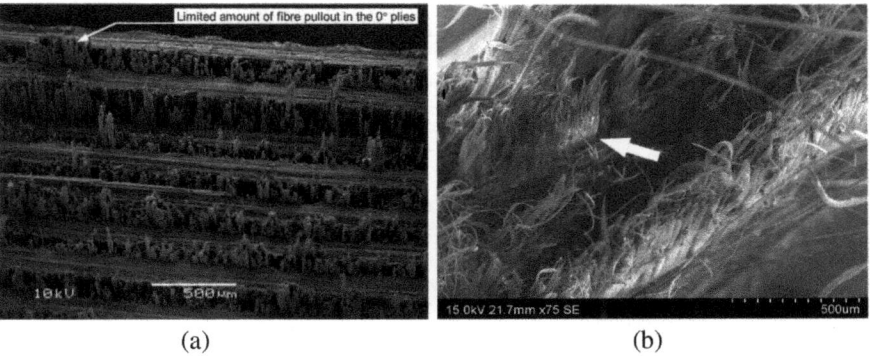

(a)   (b)

Fig. 4 Fracture surface for fibre tensile breaking mode **a** CF/Epoxy (Pinho et al. 2006) **b** Vectran/MTM57 (Syed Abdullah et al. 2018). Note the 'broomlike' features in both composites. White arrow shows fibre-matrix debonding

is poor, the degree of fibre debonding will be larger, leading to a 'broomlike' failure, shown in Fig. 4a.

For polymer fibres such as Kevlar and Dyneema, the fibre-matrix bonding strength is often relatively poor, due to the inert nature of thermoplastic materials (Herrera-Franco and Drzal 1992; Yue and Padmanabhan 1999). Furthermore, the relatively higher Poisson's ratio of polymer fibres such as Kevlar (Poisson's ratio: 0.36 (Dupont 2018)) may promote mode I debonding due to the fibre's transverse contraction under tensile loads. In addition, Hull and Clyne (1996) argued that the fracture toughness values are not restricted to pure modes I or II, but rather a mixed-mode condition. Syed Abdullah et al. (2018) and Syed Abdullah (2019) estimated the energy associated to fibre-matrix debonding on Vectran/Epoxy and found that the energy is in the range of 16.44 and 64.49 kJ/m^2. Therefore, the fracture surface of polymer fibre composites would often result in a 'broomlike' feature, shown in Fig. 4b.

Finally, considering the case of pure matrix failure, it is perhaps the first type of damage suffered by the composite upon experiencing any types of load. Depending on the composite orientation/layup, the appearance of damage may vary. For example, when a uniaxial tension load is applied normal to a fibre in a Uni-Directional (UD) composite (90°), the resulting tensile properties (strength and modulus) is usually significantly lower if compared to the result when loading parallel to the fibres (0°). This is due to several reasons. First, the mechanical properties of the matrix are significantly lower than that of the fibres (Greenhalgh 2009). Secondly, matrix-dominated failure is governed by the interfacial bonding between the fibres and the matrix, in addition to the mechanical properties itself (i.e. thermoplastics, thermosets). If the interfacial strength between the fibres and the matrix is high, translaminar fracture (fibre fracture) can develop which entails longitudinal fibre fracture. However, if the interfacial strength is low, a cohesive fracture in the matrix or fibre-matrix interfacial failure may occur, as illustrated in Fig. 5.

Fig. 5 Illustration of ply
splitting micro-mechanisms,
a cohesive fracture,
b fibre-matrix interfacial
failure, **c** translaminar
fracture

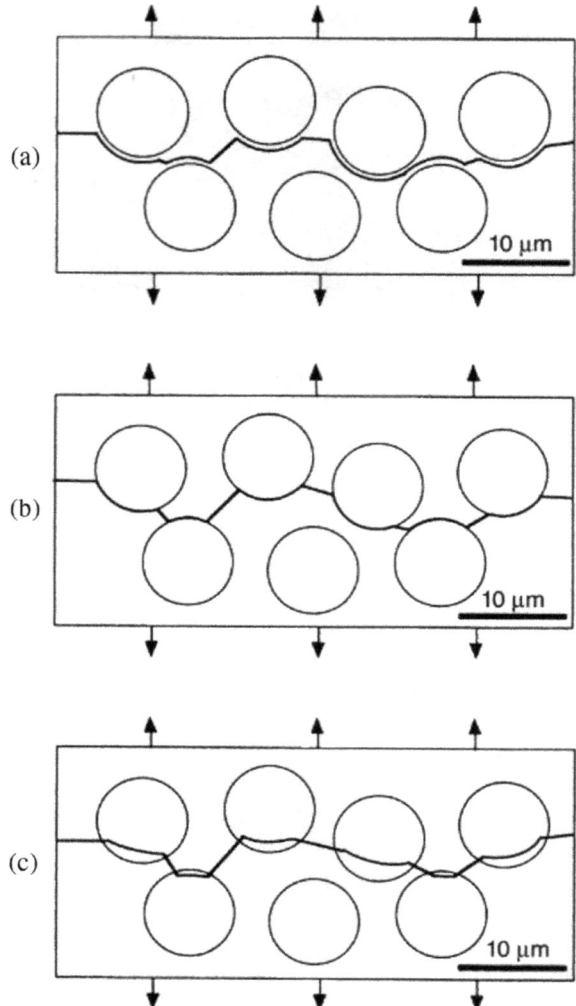

2.2 Compressive Damage

Perhaps the most important characteristic for the characterisation of damage under
compressive loading is the formation of 'kink-bands'. These bands are a result of
micro-buckling of the fibres and shear stresses which develops around the fibre-
matrix interface (Greenhalgh 2009; Pinho et al. 2006). For brittle fibre-matrix
systems, such as CF/Epoxy, kink-bands can be clearly observed on the damaged
surface cross-section, seen as fractured fibres of equal lengths propagating at an
angle, shown in Fig. 6a. Further inspection of the fibre ends reveal a clear distinction between the compression face and the tension face—the tension face having a

smoother feature compared to the compression face, whereby the surface appears rougher with small particle of debris ground into the fibre end.

For polymer fibre-based composites, failure under compression is governed by the fibre strength, which is significantly smaller than its tensile strength (less than 10% of its tensile strength and stiffness) (Greenhalgh 2009; Syed Abdullah 2019, 2021). Failure occurs almost immediately when a compressive load is applied, which resulted in local kinking of individual fibres, before propagating globally. An elastic-almost perfectly plastic response can be expected in the stress -strain curve, since the fibres do not 'fracture' but rather kink, before folding (Syed Abdullah et al. 2018; Iannucci et al. 2018; Attwood et al. 2015). In addition, the fibrillar nature of polymer fibres (Syed Abdullah et al. 2018; Greenhalgh 2009) further promotes premature failure under compressive loads.

2.3 Interlaminar (Delamination) Damage

Failure due to delamination is often a precursor on the severity of damage in the laminate. For instance, under Low Velocity Impact (LVI), a clear load drop can be observed in the time-history response, indicating a considerable degradation in the structure's integrity. In general, delamination generally develops due to an excessive out-of-plane or interlaminar stress being generated at the interfaces between adjacent plies (Greenhalgh 2009, 1998; Olsson 1992). In most cases, delamination in composite is usually a combination of mode I (opening), II (shearing), and III (tearing). However, mode III is generally neglected and a mixed-mode of I and II is usually considered.

While delamination toughness is usually determined by the matrix properties, the type of reinforcement used in the composite, the fabric architecture and layup orientation, as well as the interfacial fibre-matrix bonding is equally important due to several reasons. First, a high interfacial bonding strength is often an indication of a brittle fibre-matrix system (i.e. fibre reinforced thermosetting composites), which may have a relatively weaker delamination toughness if compared to that of fibre reinforced thermoplastic composites (Greenhalgh 1998). As the matrix toughness increases, failure tends to occur at the fibre-matrix interface, which involves a considerable amount of plastic deformation, consequently improving the delamination toughness of the composite. Another factor which may influence the delamination toughness is the choice of the reinforcing fibres. A smaller fibre diameter would promote a greater amount of fibre bridging, consequently improving the delamination toughness (Greenhalgh 2009; Hiley 1999). In addition, the fibre which has a larger tensile strain-to-failure would mean that crack propagation would be further resisted due to a tougher fibre-bridged zone in the composite. For some polymer fibres, the fibrillar nature of the fibres, Fig. 7a, would increase the degree of fibre bridging, hence considerably improving the fracture toughness. For instance, the mode I fracture toughness of Vectran/Epoxy composite was found to be 1.41 kJ/m^2 (Syed Abdullah 2019; Syed Abdullah et al. 2018, 2016), almost seven times higher than that of

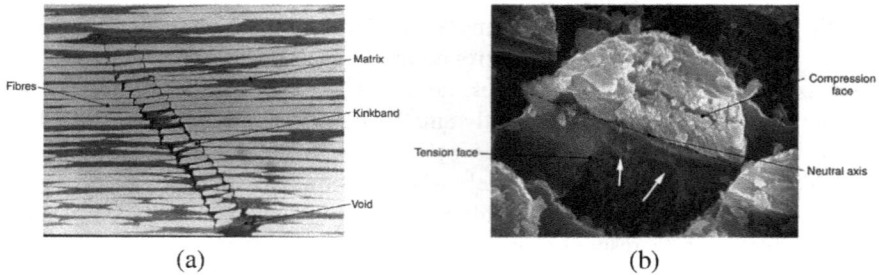

Fig. 6 Compressive failure for CF/epoxy composites **a** cross-sectional view of the kink-bands **b** close-up view of the fibre end (Greenhalgh 2009)

IM7/8552 (CF/Epoxy), which is calculated to be 0.21 kJ/m^2 (Psarras 2013). While the large fracture toughness may be influenced by the Non-Crimp Fabric (NCF) architecture, the significant degree of fibre bridging found on the fracture surface of Vectran/MTM57, shown in Fig. 7b.

Finally, the choice of fabric architecture in a composite is also a key factor which influence its delamination toughness. The enhancement in delamination toughness in woven, 2D and 3D composites is perhaps the principal reason why these architectures are selected, compared to UD. In addition to the degree of crimp and the fibre tows, fabric stitches, which are present in some of the architectures such as NCF and 3D composites act as a crack arrestor, consequently diverting the path of the crack in the direction parallel to the stitches (Greenhalgh 2009; Syed Abdullah 2019, 2021). Hence, more energy is required to further propagate the crack.

Figure 8 presents SEM images on the fracture surface of S2-Glass/Epoxy failed DCB specimens. The dotted lines shown in Fig. 7 represent riverlines, which is a characteristic feature in mode I dominated failure (Greenhalgh 2009). The interpretation of riverlines is shown in Fig. 9. Notice the change in the crack path closer to the stitch, which is then re-diverted to its initial path when moving away from the stitch. This mechanism often results in a greater energy absorption, consequently improving the composite delamination toughness.

3 Damage and Failure Under Impact

The severity of damage on a composite may vary depending on the exerted impact energy. For instance, Barely Visible Impact Damage (BVID) types are often the most serious, especially for CF/Epoxy, due to the almost linear-elastic behaviour nature of the material, which does not have any elasto-plastic component to absorb the impact energy. It may be that the composite structure has suffered extensive delamination in between plies and fibre failure, however, very little evidence of damage may be found on the external surface of the structure.

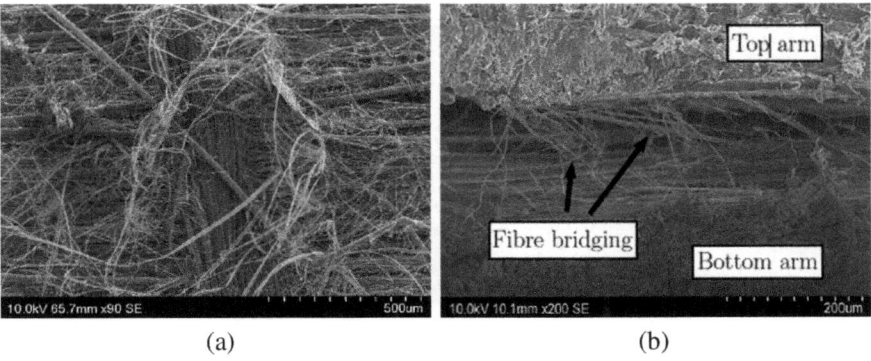

Fig. 7 Fracture surface of Vectran/MTM57 Double Cantilever Beam (DCB) specimen placed under SEM, **a** fibrils on the fracture surface, **b** Extensive fibre bridging between the top and bottom arms (Syed Abdullah 2019, 2021)

Fig. 8 Fracture surface of S2-glass/epoxy DCB specimen, **a** crack growth morphology before a stitch, **b** crack growth morphology after the stitch. Dotted lines indicate riverlines (Syed Abdullah 2019, 2021)

At low levels of impact energy, the composite may suffer minimal levels of damage. Upon initial contact between the impactor and the composite, matrix damage will appear, initiating from internal defects such as voids due to residual stresses during autoclave cure. Figure 10a and b shows the micro and meso voids which is present during composite manufacturing. In addition, certain types of fabric architecture may also introduce several 'weak' points in the composite. For instance, the formation of resin rich zones, particularly in fabric which possesses stitches such as Non-Crimp Fabrics (NCF) and 3D composites. This is shown in Fig. 10c, where resin-rich sites are present closer to the fabric stitch.

It is common that the characterization of voids are performed using non-destructive testing techniques, such as the ultrasonic C-Scan, which is based on the velocity and attenuation of the ultrasonic pulse waves through (through-transmission

Fig. 9 Illustration of how
riverlines feature develop.
Arrow indicate crack growth
(Greenhalgh 2009)

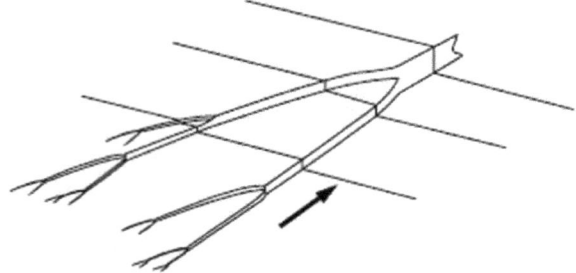

technique) or reflecting/backscattering from the back surface (pulse-echo technique) of the composite, depending on the void content and Fibre Volume Fraction (FVF) (Mehdikhani et al. 2018). Other methods, such as the microscopic techniques, are also being employed to quantify the void content in a composite, such as using the optical microscope—an example of the image produced by the optical microscope is shown in Fig. 10a and b (Greenhalgh 2009).

An increase in the impact energy would naturally result in an increase in damage levels in the composite. At higher impact loads, mode II dominated fibre-matrix debonding may occur. This is especially true for Low Velocity Impact (LVI) loads, where the global response of the composite is dominated by flexure (Davies and Olsson 2004). While improvements on the fibre-matrix interfacial properties would result in an increase in the composite in-plane mechanical properties (Totry et al. 2010; Deng and Ye 1999; Yang et al. 2019) (due to the efficient load transfer between the fibre and the matrix), a decrease in the impact performance may be observed (Bader et al. 1972; Syed Abdullah 2019, 2021). This is particularly true for brittle fibre-matrix composite systems, such as Carbon Fibre/Epoxy (CF/Epoxy) and Glass Fibre/Epoxy (GF/Epoxy), since a 'good' fibre-matrix interfacial property may lead to a brittle failure, consequently preventing the composite from further deforming to absorb the impact energy. Hence, the use of thermoplastic matrices is favoured, since it enables the matrix to deform plastically, before fully debonding from the fibre (Hull and Clyne 1996).

The coalescence of matrix cracks, which usually occur from multiple fibre-matrix debonding propagating in a ply, may initiate delamination, depending on the applied load on the structure. This can be seen in Fig. 11, where the propagation of matrix cracks in a composite resulted in an inter-ply failure (delamination) on the adjacent 0° ply.

Generally, under LVI conditions, a single large delamination frequently appears at the rearmost of the composite, along with tension matrix cracks, mainly due to flexure. It must be noted that delamination failure normally occurs only at interfaces with different fibre orientation (Richardson and Wisheart 1996; Abrate 1998), primarily induced by interlaminar shear stresses, which are enhanced by matrix cracks and ply stiffness mismatch (Davies and Olsson 2004). A typical distribution of delamination in a composite is shown in Fig. 12a and b where failure extends along the fibres of the lower plies and normally appear between plies of different orientation. Closer to the

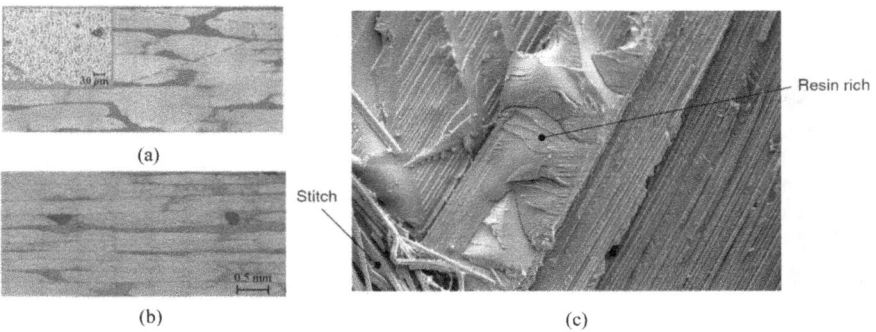

Fig. 10 Close-up images of internal defects in composites **a** micro- and **b** meso-voids inside and between tows, respectively, and **c** Resin-rich site at a high stitch tension specimen at a crimp on a Mode I fracture (HTS/RTM6) (Greenhalgh 2009; Mehdikhani et al. 2018)

front surface, local compressive and shear failure can be observed, shown in Fig. 13, due to the high contact stress between the impactor and the laminate. Conversely, a local tensile failure may be present at the rear surface of the laminate (Davies and Olsson 2004; Abrate 1998).

This is especially true for thin laminates, due to the global flexural deformation experienced by composite. In contrast, matrix cracks may be observed on the front surface for thick laminates, inducing the so-called 'pine-tree' damage pattern, shown in Fig. 13c.

The area of delamination is the largest in the laminate mid-plane, due to the shear stresses being the highest in that region. This causes a large reduction in bending stiffness, giving a significant energy-release rate (Wisnom 2012), consequently leading to a considerable degradation in the composite structural integrity. In general, a mixed-mode interlaminar failure inclined towards mode II can usually be observed in LVI events, shown in Fig. 14. Note the presence of moderately formed shear cusps around

Fig. 11 Variation of failure in a composite due to low-velocity impact loading (Davies and Olsson 2004)

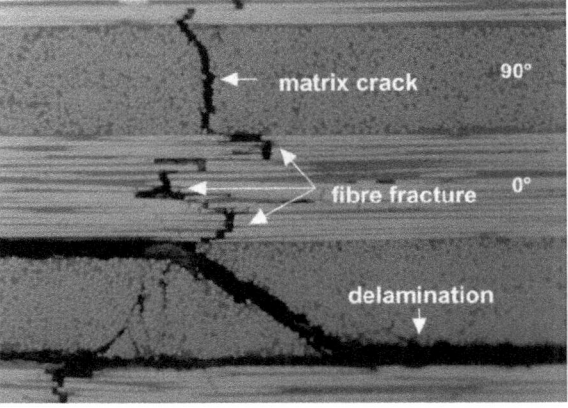

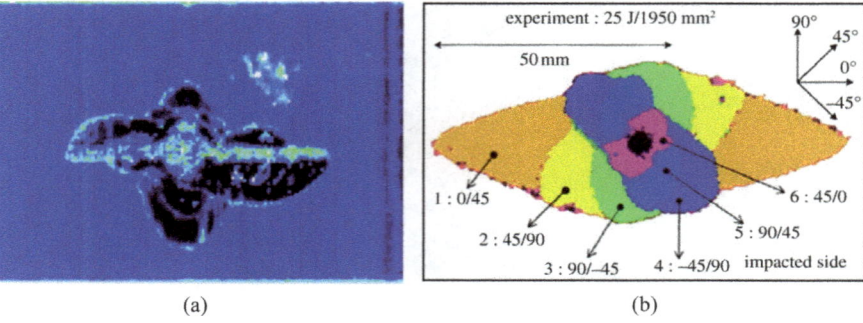

(a)   (b)

Fig. 12 Typical distribution of delamination in a laminate observed under an ultrasonic C-scan, **a** 0°/90° laminate, **b** quasi-isotropic (QI) laminate (Wisnom 2012)

the fracture surface indicating a Mode II (shear) failure has occurred. Additionally, a small presence of riverlines was observed, suggesting that a Mode I (opening) failure has taken place.

For HVI loading, stress wave effects usually dominate, inducing Mode I delamination in between plies. This is particularly true for polymer-based fibre composites such as Dyneema HB26, in which it was found that the mode II delamination were observed closer to the impact site and the laminate mid-plane, whilst away from the impact site, the Mode I component increases (Greenhalgh et al. 2013).

A further increase in the impact load would ultimately lead to fibre failure, releasing a significant amount of energy. Fibre failure often takes place closer to the end of the impact event, often resulting in laminate penetration. Therefore, a composite system based on high performance fibres (i.e. polymer-based fibres such as Kevlar and Vectran) are usually preferred due to the high fracture toughness. The high fracture toughness is an inherent characteristic of polymer-based fibres due to the considerably large plastic component in the material. A recent study by

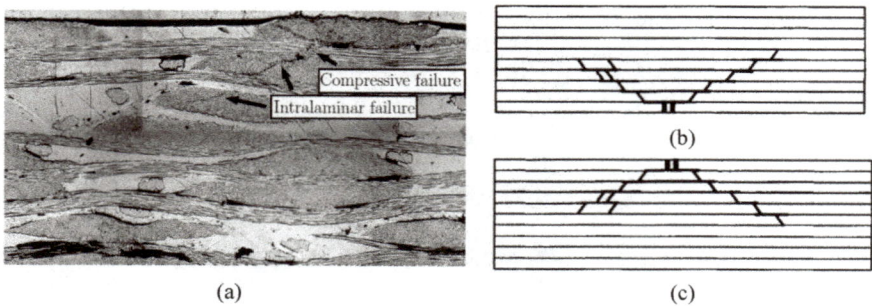

(a)   (c)

Fig. 13 a Compressive failure (kink bands) observed on the front surface of GF/Epoxy laminate under 40 J of LVI **b** reversed pine-tree damage pattern in thin laminates **c** pine-tree damage pattern in thick laminates

Fig. 14 Typical fracture surface of CF/epoxy based composite under LVI loading. Blue arrows indicate shear cusps whilst white arrows indicate riverlines

Porter et al. (2013) concluded that the fracture energy contribution from fibre failure, for most polymer fibres is around 1 kJ/m^2. Hull and Clyne (1996) argued that the energy contribution from polymer-based composites to the fracture toughness could account for up to a few kJ/m^2. Syed Abdullah et al. (2018) investigated the mode I translaminar fracture toughness of Vectran/Epoxy composite system and found that the energy associated with fibre breaking is approximately 260 kJ/m^2—significantly higher than that of CF/Epoxy and GF/Epoxy composite system (95.27% and 68.6% higher, respectively). The plastic deformation prior to failure can be visually observed under a Scanning Electron Microscope (SEM), seen as a 'ductile' drawing at the tip of the fibre, shown in Fig. 15a.

For brittle fibre composite systems, such as GF/Epoxy, little to no plastic deformation may be observed on the fibre, as shown in Fig. 15b. The surface of the fibre ends showed an almost 'planar' surface, with a clear distinction between the tension and compression face.

4 Conclusion and Future Perspectives

The complex damage types in composite laminates under impact loading has been a subject of research for many years. To date, extensive literature can be found relating to damage characterisation of composite laminates, particularly classical brittle fibre-brittle matrix types, such as CF/Epoxy and GF/Epoxy. The type of failures which are present in a composite under impact can often be deduced into a smaller component, such as tensile and compressive failures. These failure types can then be used to understand the mechanics of damage propagation (i.e. initial failure occurrence), which is unique to the material type.

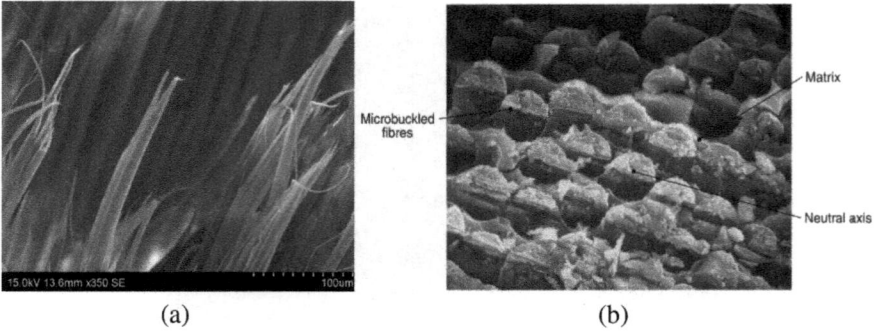

(a)   (b)

Fig. 15 SEM image for failed fibres under tension loads, **a** vectran/Epoxy showing a 'ductile' drawing at the fibre tip (Syed Abdullah 2019), **b** GF/Epoxy composite under compression (Greenhalgh 2009)

However, a consensus can be made on the type of failure which occurs in a composite under impact loading. These include (in order of failure development), matrix failure, delamination, and fibre failure. Although unique damage types will be present for some composites, for instance, fibre micro-buckling can be observed in polymer-fibre based composites (such as Kevlar and Dyneema), as opposed to fibre-kinking for brittle fibre-matrix composites (such as CF/Epoxy and GF/Epoxy) under compressive loadings. However, these types of damages can only be observed using advanced techniques such as the fractographic method using Scanning Electron Microscope (SEM).

Perhaps more importantly is the need to understand the complex damage mechanism which are present in a composite laminate under various loading types, such as impact. This is due to several reasons. First, with a comprehensive understanding, engineers could exploit the full potential of the composite, without the need to 'over-design' to ensure that the composite structure meets all safety requirements. This would result in a more cost- effective design, with minimal material consumption and fewer man-hours. Secondly, an accurate prediction of failure can be made, leading to a more reliable design since the mechanics of the composite are fully understood. Thirdly, physically based constitutive equations can be derived, which can be implemented as a material model for Finite Element Method (FEM) modelling. The material model can then be used to reduce the number of experimental tests required, leading to a reduction in the total design cost.

References

Abrate S (1998) Impact on composite structures. Cambridge University Press, Cambridge
Attwood JP, Fleck NA, Wadley HN, Deshpande VS (2015) The compressive response of ultra-high molecular weight polyethylene fibres and composites. Int J Solids Struct 71:141–155

Bader MG, Bailey J, Bell I (1972) The effect of fibre-matrix interface strength on the impact and fracture properties of carbon-fibre-reinforced epoxy resin composites. J Phys D Appl Phys 6(5):572–586

Cook J, Gordon JE (1964) A mechanism for the control of crack propagation in all-brittle systems. Proc Roy Soc A Math Phys Eng Sci 282(1391):508–520

Davies G, Olsson R (2004) Impact on composite structures. Aeronaut J 108(1089):541–563

Deng S, Ye L (1999) Influence of fibre-matrix adhesion on mechanical properties of Graphite/Epoxy composites: I. Tensile, flexure, and fatigue properties. J Reinf Plast Compos 18(11):1021–1040

Dennis N, Lyle D (2009) Bird strike damage and windshield bird strike. ATKINS, Farnham

Dupont. , 2018.Dupont (2018) Kevlar Aramid fibres technical guide. Dupont, Richmond

Greenhalgh ES (1998) Characterisation of mixed-mode delamination growth in carbon-fibre composites. Imperial College London, London

Greenhalgh ES (2009) Failure analysis and fractography of polymer composites. Woodhead Publishing Limited, Cambridge

Greenhalgh ES, Bloodworth VM, Iannucci L, Pope D (2013) Fractographic observations of Dyneema composites under ballistic impact. Compos A Appl Sci Manuf 44(1):51–62

Hedayati R, Sadighi M (2015) Bird strike—an experimental, theoretical, and numerical investigation. Woodhead Publishing, London

Hellard G (2008) Composites in Airbus: a long story of innovations and experiences. EADS Global Investors Forum, Sevilla

Herrera-Franco PJ, Drzal LT (1992) Comparison of methods for the measurement of fibre/matrix adhesion in composites. Composites 23(1):2–27

Hiley MJ (1999) Delamination between multi-directional ply interfaces in carbon-epoxy composites under static and fatigue loading. In: Fracture of polymers, composites, and adhesives. European Structural Integrity Society (ESIS), Les Diablerets, pp 61–72

Huang Y, Young RJ (1996) Interfacial micromechanics in thermoplastic and thermosetting matrix carbon fibre composites. Compos A Appl Sci Manuf 27(10):973–980

Hull D, Clyne T (1996) An Introduction to composite materials. Cambridge University Press, Cambridge

Iannucci L, Del Rosso S, Curtis PT, Duke PW (2018) Understanding the thickness effect on the tensile strength property of Dyneema®HB26 laminates. Materials 11(8):1–18

Juntikka R, Olsson R (2009) Experimental and modelling study of hail impact on composite plates. In: 17th international conference on composite materials. United Kingdom, Edinburgh

Kim H, Welch DA, Kedward KT (2003) Experimental investigation of high velocity ice impacts on woven carbon/epoxy composite panels. Compos A Appl Sci Manuf 34(1):25–41

Laffan MJ, Pinho ST, Robinson P, McMillan AJ (2012) Translaminar fracutre toughness testing of composites: a review. Polym Testing 31:481–489

Mehdikhani M, Gorbatikh L, Verpoest I, Lomov SV (2018) Voids in fibre-reinforced polymer composites: a review on their formation, characteristics, and effects on mechanical performance. J Compos Mater 53(12):1579–1669

Nguyen S, Greenhalgh ES, Graham JM, Francis A (2014) Runway debris impact threat maps for transport aircraft. Aeronaut J 108(1201):229–266

Nguyen S, Greenhalgh ES, Olsson R, Iannucci L, Curtis PT (2008) Modelling the lofting of runaway debris by aircraft tyres. J Aircr 45(5):1701–1714

Okabe T, Takeda N, Kamoshida Y, Shimizu M, Curtin W (2001) A 3D shear-lag model considering micro-damage and statistical strength prediction of unidirectional fiber-reinforced composites. Compos Sci Technol 61(12):1773–1787

Olsson R (1992) Factors influencing the interlaminar fracture toughness and its evaluation in composites. The Aeronautical Research Institute of Sweded, Bromma

Pinho ST, Robinson P, Iannucci L (2006) Fracture toughness of the tensile and compressive fibre failure modes in laminated composites. Compos Sci Technol 66(13):2069–2079

Porter D, Guan J, Vollrath F (2013) Spider silk: super material or thin fibre. Adv Mater 25(9):1275–1279

Psarras S (2013) Investigating failure in composite stiffener run-outs. Imperial College London, London

Richardson MO, Wisheart WJ (1996) Review of low-velocity impact properties of composite materials. Compos A Appl Sci Manuf 27(12):1123–1131

Roseler WG, Sarh B, Kismarton MU (2007) Composite Structures: the first 100 years. In: 16th international conference on composite materials. Japan, Kyoto

Rosen BW (1964) Tensile failure of fibrous composites. AIAA J 2(11):1985–1991

Syed Abdullah SIB (2019) The impact behaviour of high performance fibre composites. Imperial College London, London

Syed Abdullah SIB, Iannucci L, Greenhalgh ES, Ahmad Z (2021) The impact performance of Vectran/Epoxy composite laminates with a novel non-crimp fabric architecture. Compos Struct 265:1–17

Syed Abdullah SIB, Iannucci L, Greenhalgh ES (2016) An experimental study of the mode i interlaminar fracture toughness of vectran/epoxy composites. In: European conference on composite materials. Germany, Munich

Syed Abdullah SIB, Iannucci L, Greenhalgh ES (2018) On the translaminar fracture toughness of Vectran/epoxy composite material. Compos Struct 202:566–577

Totry E, Molina-Aldareguia J, Gonzalez C, Llorca J (2010) Effect of fiber, matrix and interface properties on the in-plane shear deformation of carbon-fiber reinforced composites. Compos Sci Technol 70(6):970–980

Wang WT, Yu H, Potter K, Kim BC (2018) The interlaminar fracture toughness of carbon/epoxy laminates interleaved with polyamide particle layers. In: Proceedings of the European conference of composite materials (ECCM 18). Athens, Greece

Wisnom MR (2012) The role of delamination in failure of fibre-reinforced composites. Philos Trans Roy Soc A Math Phys Eng Sci 370(1965):1850–1870

Yang G, Park M, Park S (2019) Recent progresses of fabrication and characterization of fibers-reinforced composites: a review. Compos Commun 14:34–42

Yue CY, Padmanabhan K (1999) Interfacial studies on surface modified Kevlar fibre/epoxy matrix composites. Compos B Eng 30(2):205–217

Zhuang L, Talreja R, Varna J (2016) Tensile failure of unidirectional composites from a local fracture plane. Compos Sci Technol 133:119–127

Zweben C (1968) Tensile failure of fibre composites. AIAA J 6(12):2325–2331

Low Velocity Impact, Ultrasonic C-Scan and Compression After Impact of Kenaf/Jute Hybrid Composites

K. Tabrej, M. T. H. Sultan, M. Jawaid, A. U. M. Shah, and S. Sani

Abstract The current research will discuss on the low velocity impact properties of kenaf/jute/kenaf hybrid composites treated with sodium hydroxide solution. The after impact damage properties will be observed through ultrasonic C-scan and compression after impact analyses. Kenaf/jute/kenaf sequence with 30 wt% fibre loading was laid up in epoxy matrix and cured in room temperature. The hybrid composites can withstand up to 30 J impact energy without full penetration, but appeared to be severely damaged. The visual inspection through the naked eyes clearly shown the cracks propagated towards the sides of the samples. However, limitations occurred through the C-scan technique, as the damages captured were not fully in parallel to the visual inspection. This is due to the properties of natural fibres. The compression after impact testing showed a maximum force of 42 kN needed to break the samples, impacted with low impact energy. The maximum compression force reduced as the impact energy increased.

Keywords Natural fibers · Kenaf · Jute · Hybrid composites · Low velocity impact

K. Tabrej · M. T. H. Sultan (✉) · A. U. M. Shah
Department of Aerospace Engineering, Faculty of Engineering, Universiti Putra Malaysia, 43400 UPM Serdang, Selangor Darul Ehsan, Malaysia
e-mail: thariq@upm.edu.my

M. T. H. Sultan · M. Jawaid · A. U. M. Shah
Laboratory of Biocomposite Technology, Institute of Tropical Forestry and Forest Products (INTROP), Universiti Putra Malaysia, 43400 UPM Serdang, Selangor Darul Ehsan, Malaysia

M. T. H. Sultan
Aerospace Malaysia Innovation Centre (944751-A), Prime Minister's Department, MIGHT Partnership Hub, Jalan Impact, 63000 Cyberjaya, Selangor Darul Ehsan, Malaysia

S. Sani
Industrial Technology Division, Malaysian Nuclear Agency, 43000 Kajang, Selangor, Malaysia

© Springer Nature Singapore Pte Ltd. 2021
M. T. H. Sultan et al. (eds.), *Impact Studies of Composite Materials*, Composites Science and Technology, https://doi.org/10.1007/978-981-16-1323-4_6

1   Introduction

Kenaf and jute natural fibers are much famous now as an alternative to synthetic (Barari et al. 2016), and both fibers have some good and bad advantages such as kenaf and jute are very much eco-friendly, the cheap cost to making products etc. (Omrani et al. 2016). and the disadvantages such as poor compatibilities in petroleum-based polymers compared to synthetic fibre reinforcements (Aji et al. 2013). The researcher is trying to find out the way to the enhancement of properties of natural kenaf/jute hybrid composites (Ferdous and Hossain 2017). The hydrophilicity decreases the adhesion to the polymer matrix, thus decreasing the performance of the obtained composites (Ku et al. 2011). Chemical modifications on natural fibers can decrease its hydrophilic properties and improve the compatibilities between the matrix and fibers (Pickering et al. 2016). Previous studies suggested chemical modification on natural fibres such as alkali and silane treatment (John and Anandjiwala 2008). In this study, sodium hydroxide (NaOH) fibers treatment was applied to improve the compatibilities of kenaf/jute hybrid composites.

Hybrid is one of the alternative techniques to improve the disadvantages of single type composites. There are natural-natural, natural-synthetic and synthetic-synthetic types of hybrid composites (Gholizadeh 2018; Luo 2016). To reduce the percentage of synthetic materials in composites, natural-natural hybrid is the best alternative (Jothibasu et al. 2018). Various combinations of natural-natural hybrid composites were reported to study their mechanical, thermal, physical and impact properties (Mansor et al. 2013). Studies on kenaf and jute as individual reinforcement in polymer matrix composites had been initiated by some researcher few years back. However, the hybrid of these two fibres was not yet reported to date. This is one of the gap that can be explored to suggest more potential applications of the hybrid composites (Ali et al. 2017; Selver et al. 2016).

Damages in composites can be analysed through destructive and non-destructive methods. Ultrasonic C-scan is one of the promising techniques to evaluate damages in laminated composites. However, its use in natural fibre composites is very limited. The concept of relative attenuation ultrasonic waves is applied in this technique. A transducer is used to scan the materials with the help of water as the transmission medium between both surfaces. Comparison between the relative attenuation and the operator-set level will results in colored images to display the damages of composites (Rahman et al. 2015). Besides matrix cracking, ultrasonic C-scan is also able to detect voids and delamination of composites (Ibrahim 2016). The ability of this technique to detect interlaminar damages in composites suggests its use in various fields (Dubary et al. 2018). This method is mostly suggested for the quality control in the production line of product development (Segreto et al. 2018).

Compared to tension and bending forces, low velocity impact cause higher risk to the strength and durability of composite structures. This is due to the sudden force applied in a very short time on the structure, which results in unpredictable damages (Razali et al. 2019). Due to that, compression after impact (CAI) testing is an excellent technique suggested to examine the residual strength of the impacted composite. However, this method is a destructive evaluation method, which needs to be done during the development and characterization stage of the materials and not applied on the complex structure of respective applications (Safri et al. 2019). Previous studies suggested a good correlation between the level of impact energy and the residual compressive strength of sugar palm/glass hybrid composites, which the residual strength reduced as the impact energy increases (Safri et al. 2019). Similar trend was also observed on the kenaf/glass hybrid composites using woven type fibres (Ismail et al. 2019). The breaking point during the compression after impact testing generally more concentrated at the centre, where the samples were impacted during the low velocity impact (Shahzad 2019). The severity of damage highly affects the residual strength and the damage pattern during the compression after impact testing (Sasikumar et al. 2019). Samples with bigger damage area, but less depth of damage are easier to be compressed compared to samples having smaller damage area with thicker depth (Khan et al. 2018a, b).

The current study aims to analyse the low velocity impact properties of treated kenaf/jute hybrid composites. The impact damages will be observed through visual inspection and ultrasonic C-scan, while the residual strength will be analysed through compression after impact testing.

2 Methodology

2.1 Materials

In the current study, woven mat kenaf and jute fibres were supplied by Indersen Shamlal Pvt. Ltd. India while Epoxamite 100 with 102 medium hardeners, brand Smooth-On, were purchased from Mecha Solve Engineering Sdn. Bhd, Selangor, Malaysia. The epoxy and hardener were mixed at a ratio of 3:1 as per datasheet provided. Table 1 showed the properties of Kenaf and Jute fibres.

Table 1 Properties of kenaf and Jute Fibers (Khan et al. 2018a, b)

Properties	Kenaf fiber	Jute fiber
Density	1.4	1.3
Tensile's strength	930	393–773
Young's modulus	20	26.5
Elongation at break (%)	1.6	1.5–1.8
Cellulose content (%)	53–11	58–63
Hemicellulose content (%)	15–19	12%
Lignin content (%)	05–11	12–14

2.2 Preparation of Composites

All the woven kenaf and jute fibres used in this study were immersed in a 6% NaOH solution for 2 h a room temperature. The fibres were then washed with running tap water and oven-dried at 80 °C for 24 h. The kenaf/jute hybrid composites were fabricated in a total of 9 layers with three layers each at the sequence of KKK/JJJ/KKK. This hybrid composites were fabricated using hand lay-up method and cured at room temperature for 24 h.

3 Low Velocity Impact Test

Low velocity impact testing was performed using the Imatek IM1-C Drop Weight Testing Machine, at the Laboratory of Aerospace Structure, UPM, as shown in Fig. 1. All results obtained from the low velocity impact tests were generated using Imatek's software. This computer software works together with impact tester and data acquisition system. Five impact energy levels of 10, 15, 20, 25, and 30 J, were applied in this study.

These energy levels (in Joules) were selected based on the preliminary tests conducted to obtain the range of energy reasonable for the whole actual impact testing. The energy levels more than 30 J lead to ultimately break the specimen. Low-velocity impact test was completed on the kenaf/jute hybrid composites with five repetitions of the impact energy of each sample. The drop mass of the hemispherical steel impactor of 5.101 kg and the tip diameter of 16 mm remains constant during the impact testing. All tests were performed on the specimens with sizes of 100 mm × 150 mm × 8.30 mm, according to ASTM (D7136, 2015). Equation (1) was applied to obtain the height of impactor based on the required impact energy.

$$\text{Imapact energy, } E_i = mgh \tag{1}$$

where,

m Total mass of the impactor, 5.101 kg
g Gravitational acceleration, 9.81 m/s^2
h Height of the impactor.

4 Ultrasonic C-Scan

The damage area of all the impacted samples were observed using the ultrasonic C-scan. The procedures were carried out at Malaysian Nuclear Agency, Bangi, Selangor using the R-Theta Arm scanner and RθScan software. Transducer with capacity of

Fig. 1 Drop test rig
instrument for low-velocity
impact testing

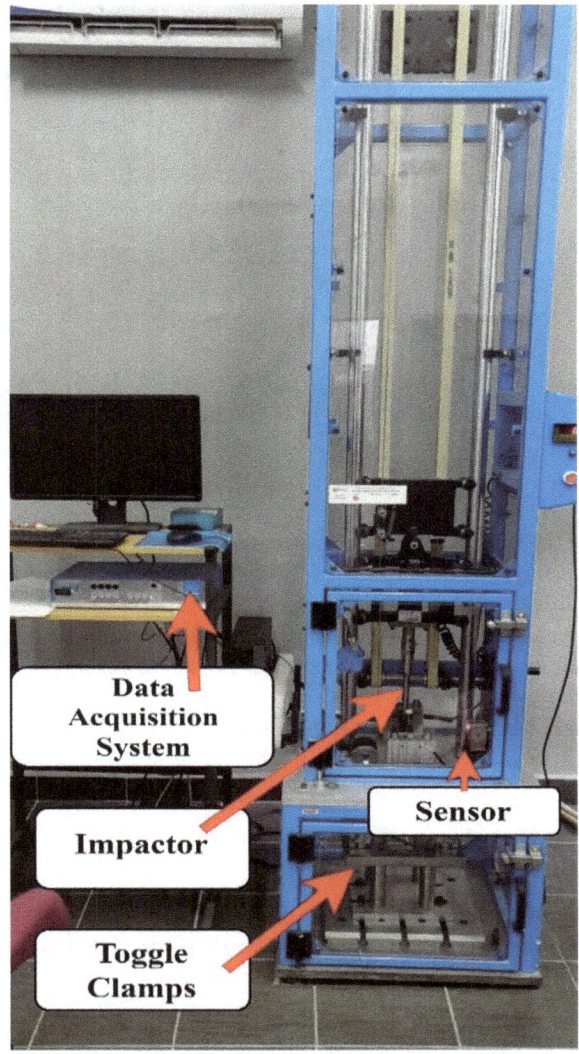

1 MHz was used, together with SONATEST gel as couplant agent. The procedures
were shown in Fig. 2.

5 Compression After Impact

Compression after impact (CAI) testing was conducted at Universiti Teknologi
Malaysia (UTM), Kuala Lumpur. The test was performed according to ASTM

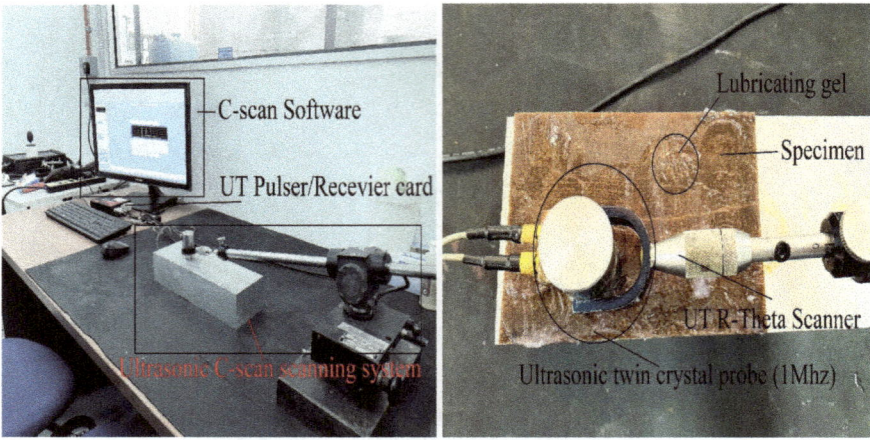

Fig. 2 The ultrasonic C-scan set up

(D7137, 2012) standard for assessing the compressive strength after the impact for a sample size of 100×150 mm. The CAI investigation was carried out by using the Universal Testing Machine (UTM) with 300 kN maximum load capacity. The typical crosshead speed displacement rate was 0.5 mm/min. Figure 3 shows the experimental set up for the compression after impact testing. A specific jig was used to hold the samples in upright position while the compressive load was applied equally the top side of the impacted samples until it breaks.

6 Results and Discussions

In this study, the analysis of low velocity impact properties were discussed in two different curves, which are force–displacement and force–time curves respectively.

6.1 Force Versus Displacement

The severity of damage on specimens during the low velocity impact test can be understood from the graph of force–displacement as shown in Fig. 4. The impactor movement and the deformation of the specimens impacted on the surface during interaction with the impactor are described in terms of the displacement. All samples showed the same trend at 10–30 J. As in Fig. 4, the closed curve obtained from the graph of force–displacement specified the specimens are being tested for the incomplete penetration damage on the test specimens. Indirectly, this clarified that low velocity full impactor penetration produced an open curve in the force–displacement

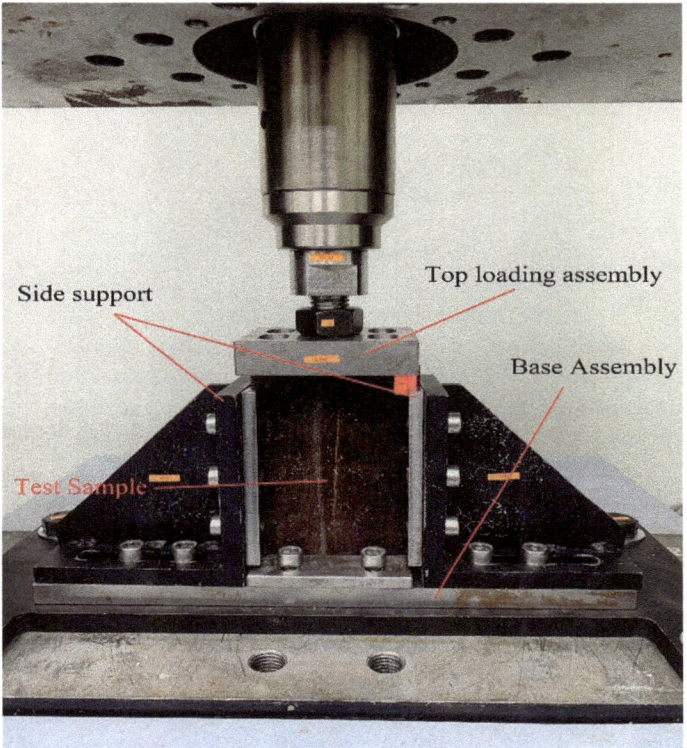

Fig. 3 The compression after impact test set up

Fig. 4 Force versus displacement graph for woven kenaf/jute hybrid composites

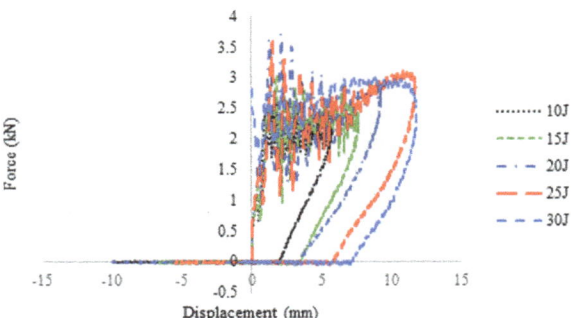

graph. The upward and downward sections of the closed curve described the loading and unloading region respectively. The upward component also delivered the data about the sufficient bending stiffness on the sample (Salleh et al. 2013).

Moreover, another additional observation was the energy absorbed by the specimens. This value is significant to the area under this force–displacement graph. The

Fig. 5 Force versus time
graph for woven kenaf/jute
hybrid composites

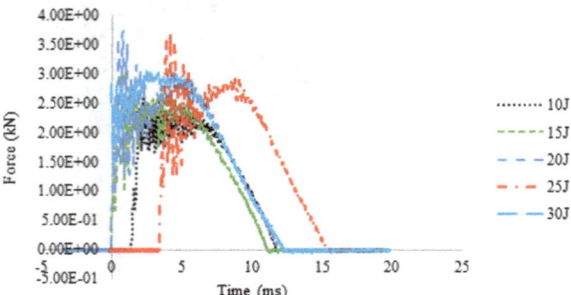

kinetic energy transferred from the impactor to the specimens was absorbed in the form of damage initiations. In this study, kenaf as the outer layer act as protector to the jute in the inner layer. The displacement showed increasing value with the increase of impact energy levels (Campo 2008). The increase in displacement had increased the area under the graph, which in parallel to the more severe damage occurred for samples impacted with higher impact energy (Maleque et al. 2012).

6.2 Force Versus Time

The force–time curves in Fig. 5 showed the continuous rough and unsymmetrical shape from the beginning of sample fractured until damage (Yaghoobi and Fereidoon 2018). The area where the sample begins to fracture was indicated before the peak load and the propagation of the fracture was indicated beyond the peak load. For the kenaf/jute composites, the loading curve increased proportionately until a sawtooth-like curve was established at peak force before unloading occurred. From the graph, the force shifted in the curved slope beyond the first peak force fell due to the reduction in material stiffness for the kenaf/jute hybrid composite (Belingardi and Vadori 2002).

Initially, the kenaf/jute hybrid composite elastically behaved with minor failures like micro-cracking. Then, significant damaged occurred due to high force. The loading rate increased according to energy levels and significantly increased the force.

6.3 Ultrasonic C-Scan

Ultrasonic C-scan provided an appropriate analysis of the damage area of composites. In this study, the images from the ultrasonic C-scan were compared with the visual inspection from the naked eyes. Figure 7 shows the comparisons of damage area of

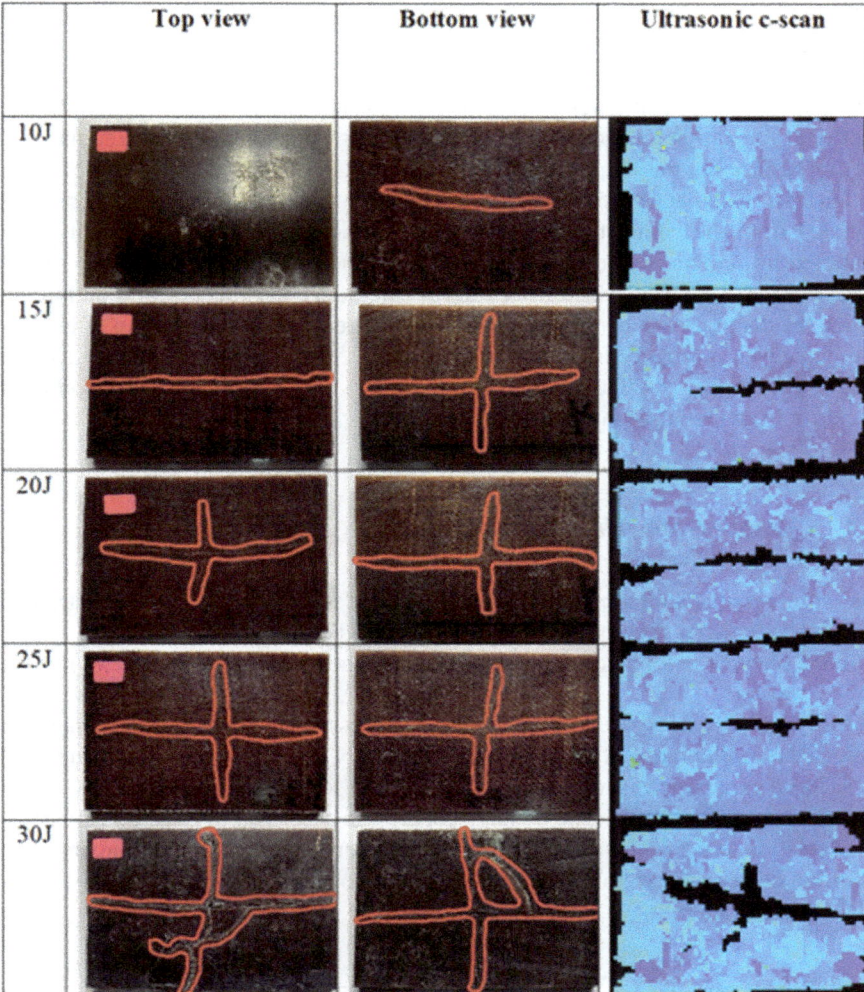

	Top view	Bottom view	Ultrasonic c-scan
10J			
15J			
20J			
25J			
30J			

Fig. 7 Comparison of damage observations through visual inspection and images from ultrasonic C-scan

kenaf/jute hybrid composites impacted with respective energy levels of 10, 15, 20, 25, and 30 J.

All the five samples repetition for each energy level were scanned and analysed. However, only the most visible damages were compared in Fig. 7. In overall, the images from the C-scan technique did not showed a reliable data on the damage area. This is due to the natural fibres used as the reinforcement in composites, which absorbed the waves used in this technique. It can be suggested in the future, that the scanning procedure to be done several times on the surfaces of composites to collect more data and produce better images.

It can be seen the damage at the bottom surface of the sample impacted with 10 J impact energy could not be detected at all using the ultrasonic C-scan. Image produced for sample impacted at 15 J impact energy also displayed the damage on the top surface but missing for the bottom surface (Robinson and Davies 1992).

In terms of damage propagation in composites, it can be seen that matrix cracking initiated from the point of impact towards the sides of samples in a line and not scattered in numerous directions. This trend is one of the effects of using woven type fibres in composites. The woven fibres can reduce the chances of total failure, which the damages can be scattered in different directions from the point of impact. This condition normally happened on short fibre composites or random oriented long fibres composites. These two types of composites also have lower visibility of damages from the ultrasonic C-scan technique.

6.4   Compression After Impact (CAI)

The compression after impact (CAI) test is used to predict the residual strength of composites after they had been impacted with respective impact energy levels. Figure 8 shows the force–displacement graph of the samples for compression after impact testing.

In overall, it can be seen that the peak compression force to break the samples were in the similar range of approximately 40–42 kN, except for the sample impacted at 30 J as it experienced severe damages and nearly to total failure during the impact (Ghelli and Minak, 2011). The damage patterns as described in Sect. 6.3 showed a good relation to the CAI analysis. All samples experienced matrix cracking initiated from the center, which promotes failure of composites under compressive load at almost similar force.

Fig. 8 Compression force vs displacement of kenaf/jute hybrid composite

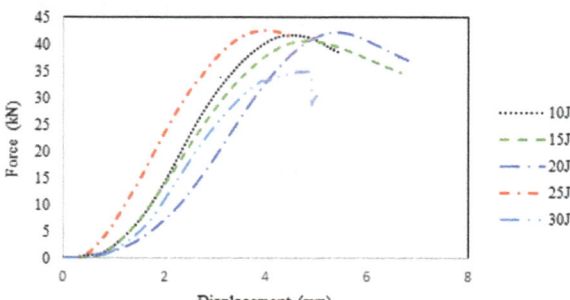

7 Conclusions

The current study highlights the low velocity impact properties of woven kenaf/jute hybrid composites, their damage analyses and residual strength after impact. The use of woven type fibres results in more consistent data in all the analyses conducted. The ability of these natural-natural hybrid composites to withstand high impact energy of 30 J can be contributed from the type of woven fibres and the NaOH treatment applied on the fibres to improve the compatibility in polymers. In overall, this study may contribute to a better potential development of natural fibre composites, without having synthetic fibres as hybrid. The key findings from the current study can be suggested as follows:

i.   Kenaf/jute hybrid composites in the sequence of KKK/JJJ/KKK in epoxy matrix with a total thickness of approximately 8 mm can withstand maximum impact energy of 30 J through the drop test rig method.
ii.   The KKK/JJJ/KKK hybrid composites did not show visible damage on the top surface at 10 J impact energy.
iii. The use of natural fibres in polymer composites limits the data collected in the ultrasonic C-scan technique, thus produced non-reliable images of damage in comparison to the visual inspection from the naked eyes.
iv.   The damage patterns observed after the impact testing can be correlate with the data acquired from the compression after impact testing.
v.   Increase in impact energy will increase the severity of damages, thus reduce the residual strength of composites.

Acknowledgements This work was supported by UPM under GPB grant, 9668200. The authors would like to express their gratitude and sincere appreciation to the Department of Aerospace Engineering, Universiti Putra Malaysia, and Laboratory of Bio-composite Technology, Institute of Tropical Forestry and Forest Products (INTROP), Universiti Putra Malaysia, UPM (HICOE) for the close collaboration in this research.

Conflict of Interests The authors declare that there is no conflict of interests regarding the publication of this paper.

References

Aji IS, Zainudin ES, Abdan K, Sapuan SM, Khairul MD (2013) Mechanical properties and water absorption behavior of hybridized kenaf/pineapple leaf fibre-reinforced high-density polyethylene composite. J Compos Mater 47(8):979–990
Ali MAEM, Soliman AM, Nehdi ML (2017) Hybrid-fiber reinforced engineered cementitious composite under tensile and impact loading. Mater Des 117:139–149
Barari B, Omrani E, Moghadam AD, Menezes PL, Pillai KM, Rohatgi PK (2016) Mechanical, physical and tribological characterization of nano-cellulose fibers reinforced bio-epoxy composites: an attempt to fabricate and scale the 'Green' composite. Carbohydr Polym 147:282–293

Belingardi G, Vadori R (2002) Low velocity impact tests of laminate glass-fiber-epoxy matrix composite material plates. Int J Impact Eng 27(2):213–229

Campo EA (2008) Selection of polymeric materials: how to select design properties from different standards. William Andrew

D7136, ASTM (2015) Standard test method for measuring the damage resistance of a fiber-reinforced polymer matrix composite to a drop-weight impact event.https://doi.org/10.1520/d7136_d7136m-15

D7137, ASTM (2012) Standard test method for compressive residual strength properties of damaged polymer matrix composite plates. https://doi.org/10.1520/d7137_d7137m-12

Dubary N, Bouvet C, Rivallant S, Ratsifandrihana L (2018) Damage tolerance of an impacted composite laminate. Compos Struct 206:261–271

Ferdous S, Hossain MS (2017) Natural fibre composite (NFC): new gateway for jute, kenaf and allied fibres in automobiles and infrastructure sector. World J Res Rev 5(3)

Ghelli D, Minak G (2011) Low velocity impact and compression after impact tests on thin carbon/epoxy laminates. Compos Part B Eng 42(7):2067–2079

Gholizadeh S (2018) A review of impact behaviour in composite materials. In: Paper presented at the proceedings of Academicsera 38th international conference

Ibrahim ME (2016) Nondestructive testing and structural health monitoring of marine composite structures. In: Marine applications of advanced fibre-reinforced composites. Elsevier, pp 147–183

Ismail MF, Sultan MTH, Hamdan A, Shah AUM, Jawaid M (2019) Low velocity impact behaviour and post-impact characteristics of kenaf/glass hybrid composites with various weight ratios. J Mater Res Technol 8(3):2662–2673

John MJ, Anandjiwala RD (2008) Recent developments in chemical modification and characterization of natural fiber-reinforced composites. Polym Compos 29(2):187–207

Jothibasu S, Mohanamurugan S, Vijay R, Lenin Singaravelu D, Vinod A, Sanjay MR (2018) Investigation on the mechanical behavior of areca sheath fibers/jute fibers/glass fabrics reinforced hybrid composite for light weight applications. J Indus Text 1528083718804207

Khan MZ, Nazir H, Yifei H, Hong S, Ahmed FU, Liu K (2018a) Mechanical properties of ambient cured high-strength plain and hybrid fiber reinforced geopolymer composites from triaxial compressive tests. Constr Build Mater 185:338–353

Khan T, Sultan H, Bin MT, Ariffin AH (2018b) The challenges of natural fiber in manufacturing, material selection, and technology application: a review. J Reinf Plast Compos 37(11):770–779. https://doi.org/10.1177/0731684418756762

Ku H, Wang H, Pattarachaiyakoop N, Trada M (2011) A review on the tensile properties of natural fiber reinforced polymer composites. Compos Part B Eng 42(4):856–873

Luo T (2016) Mechanical behavior of layered foam composite panels subjected to low velocity impact events

Maleque MA, Afdzaluddin A, Iqbal M (2012) Flexural and impact properties of kenaf-glass hybrid composite. Paper presented at the Advanced Materials Research

Mansor MR, Sapuan SM, Zainudin ES, Nuraini AA, Hambali A (2013) Hybrid natural and glass fibers reinforced polymer composites material selection using analytical hierarchy process for automotive brake lever design. Mater Des 51:484–492

Omrani E, Menezes PL, Rohatgi PK (2016) State of the art on tribological behavior of polymer matrix composites reinforced with natural fibers in the green materials world. Eng Sci Technol Int J 19(2):717–736

Pickering KL, Efendy MGA, Le TM (2016) A review of recent developments in natural fibre composites and their mechanical performance. Compos Part a Appl Sci Manuf 83:98–112

Rahman MM, Hosur M, Hsiao K-T, Wallace L, Jeelani S (2015) Low velocity impact properties of carbon nanofibers integrated carbon fiber/epoxy hybrid composites manufactured by OOA–VBO process. Compos Struct 120:32–40

Razali N, Sultan MT, Jawaid M (2019) Impact damage analysis of hybrid composite materials. In: Durability and life prediction in biocomposites, fibre-reinforced composites and hybrid composites. Elsevier, pp 121–132

Robinson P, Davies GAO (1992) Impactor mass and specimen geometry effects in low velocity impact of laminated composites. Int J Impact Eng 12(2):189–207

Safri SN, Sultan MT, Jawaid M, Majid MA (2019) Analysis of dynamic mechanical, low-velocity impact and compression after impact behaviour of benzoyl treated sugar palm/glass/epoxy composites. Compos Struct 226:111308

Salleh Z, Hyie KM, Berhan MN, Taib YM, Hassan MK, Isaac DH (2013) Effect of low impact energy on kenaf composite and kenaf/fiberglass hybrid composite laminates. Paper presented at the Applied Mechanics and Materials

Sasikumar A, Trias D, Costa J, Blanco N, Orr J, Linde P (2019) Effect of ply thickness and ply level hybridization on the compression after impact strength of thin laminates. Compos Part a Appl Sci Manuf 121:232–243

Segreto T, Teti R, Lopresto V (2018) Non-destructive testing of low-velocity impacted composite material laminates through ultrasonic inspection methods. In: Characterizations of some composite materials, IntechOpen

Selver E, Potluri P, Hogg P, Soutis C (2016) Impact damage tolerance of thermoset composites reinforced with hybrid commingled yarns. Compos Part B Eng 91:522–538

Shahzad A (2019) Investigation into fatigue strength of natural/synthetic fiber-based composite materials. In: Mechanical and physical testing of biocomposites, fibre-reinforced composites and hybrid composites. Elsevier, pp 215–239

Yaghoobi H, Fereidoon A (2018) An experimental investigation and optimization on the impact strength of kenaf fiber biocomposite: application of response surface methodology. Polym Bull 75(8):3283–3309

Impact and Post-impact Analysis on Engineered Composites

Sunil Chandrakant Joshi and Yi Di Boon

Abstract One of the main downsides of fiber reinforced polymer composite laminates is their susceptibility to impact damage. Many different methods have been
devised to either modify or engineer the composites to improve their impact resistance and tolerance. In this chapter, three types of engineered composites and
their impact performances are presented and discussed. The engineered composites are (i) composites with core–shell polymer particles, (ii) composites with carbon
nanotubes/nanofibers and (iii) composites with thermoplastic film interleaves. Core–
shell polymer particles absorb large amounts of energy during impact, thus limit
the damage done to the composite laminate. Carbon nanotubes and nanofibers have
excellent stiffness and strength, making them suitable for reinforcing the interlaminar
regions of composite laminates, leading to improved impact resistance. For composites with thermoplastic film interleaves, thermoplastic films with high intrinsic toughness modify the interlaminar regions of the composite laminate and enhance the
overall toughness. The three types of engineered composites exhibit varying levels
of improved impact performances compared to the unmodified composites. For each
type of engineered composite, its impact response, the strengthening mechanisms
and factors affecting its impact performance are deliberated.

Keywords Low velocity impact · Engineered composites · Core shell polymer ·
Carbon nanotube/nanofiber · Film interleaving

1 Introduction

Fiber reinforced polymer (FRP) composites are popular materials used in structural
applications due to their very high specific strengths. One of the main weaknesses of
FRP composites is their susceptibility to impact damage, even at low energy levels.
Low energy impact coming from tool drops, collision with debris or other events can
result in barely visible and difficult-to-detect internal damage to FRP composites,

S. C. Joshi (✉) · Y. Di Boon
School of Mechanical and Aerospace Engineering, Nanyang Technological University, 50,
Nanyang Avenue, Jurong West 639798, Singapore
e-mail: mscjoshi@ntu.edu

© Springer Nature Singapore Pte Ltd. 2021   87
M. T. H. Sultan et al. (eds.), *Impact Studies of Composite Materials*, Composites Science
and Technology, https://doi.org/10.1007/978-981-16-1323-4_7

which affect their load carrying capabilities greatly (Abrate 1991). Therefore, it is important to study the effects of impact and ways to improve impact resistance of FRP composites.

1.1 Impact Damage

Impact events on FRP composite laminates can be categorized as low energy impact, ballistic impact and hypervelocity impact. Ballistic impact is associated with military applications, while hypervelocity impact refers to impact at very high speeds (above 1 km/s) relevant for spacecraft design and analysis (Abrate 1991). For composite materials in structural applications, low energy impact events are more common and will be the focus of the chapter. Low energy or low velocity impact can lead to an internal damage in a laminate, but often produce only a slight indentation to the outer surface of the laminate (termed as barely visible impact damage) (Abrate 1991; Kumar and Rai 1993).

The damage in FRP composites due to impact include fiber breakage, matrix cracking and delamination modes (Abrate 1991; Khan and Kim 2011). For fiber damage modes, impact damage is largely affected by the strain energy absorbing capability of the fiber reinforcement. Fiber reinforcement with high strain to failure leads to composites with improved impact resistance (Cantwell and Morton 1991). The delamination and matrix cracking damage modes are largely dependent on the mechanical properties of the polymer matrix. Many composites in structural applications have brittle matrices (such as epoxy), leading to low resistance to delamination and impact damage (Sela and Ishai 1989).

For low velocity impact, experimental tests that can be performed on FRP composites include the Charpy test, the Izod test and the drop weight impact test. The different test methods have been detailed by Cantwell and Morton (Cantwell and Morton 1991). For the studies discussed in this chapter, the drop weight impact test is used. Figure 1 shows typical load and energy response curves obtained from a drop weight impact test. From the load response curve, damage initiation can be determined by identifying the first load drop occurrence. The load at damage initiation is denoted by L_{init}. The peak load, L_{peak}, can be determined from the highest point of the load response curve. This is the impact load that the composite can withstand before undergoing severe damage (Rahman et al. 2015).

The characteristics determined from the energy response curve are also indicated in Fig. 1. For impact tests without perforation, some energy is transferred back to the impactor, causing the impactor to rebound. This is the elastic portion of the stored energy in the specimen, E_{elas}. The remaining energy, E_{abs}, is absorbed by the specimen, resulting in damage such as delamination (Walker et al. 2002). For impact tests leading to perforation, there is no E_{elas} (impactor does not rebound) and the energy absorbed is also termed the perforation energy. Besides the load and energy response curves, the damage area due to the impact can also be used as a measure

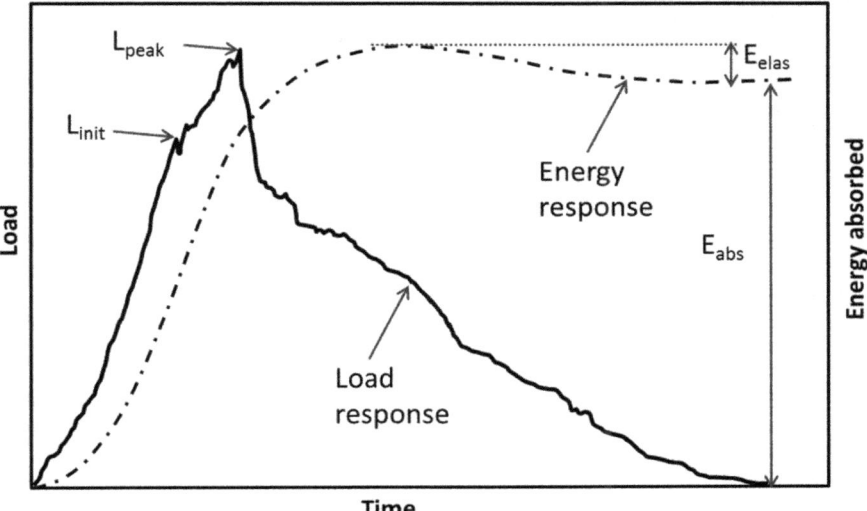

Fig. 1 Typical load and energy vs time responses of FRP composite laminate subjected to low energy impact

of impact resistance. The damage area can be determined using methods such as ultrasonic C-scan.

1.2 Residual Strength

Besides impact resistance, it is also important to study the impact damage tolerance of an FRP composite. Damage tolerance is "the ability of a material to survive a specific amount of damage" (Khan and Kim 2011). Impact damage tolerance can be studied by measuring the residual strength of an FRP composite after being subjected to impact damage.

Compression-after-impact (CAI) tests are used to study the residual compression strength. CAI tests can be performed following the procedure detailed in the ASTM D7137 standard. A fixture to support and guide the test specimen is required to prevent bending and buckling in the specimen during the CAI test. Researchers have also used flexural-after-impact (FAI) tests to study the impact damage tolerance of FRP composites (He et al. 2019; Santiuste et al. 2010). There are currently no specific standards for FAI tests. Three or four point flexural tests (as detailed in standards such as ASTM D790 and D6272) can be carried out on specimens subjected to impact damage to determine the residual flexural strength and stiffness. Strength and stiffness retention factors can be calculated as the ratio of (residual property)/(property of undamaged specimen) (Abrate 1991).

1.3   Engineered Composites

Researchers have studied various methods to engineer or modify FRP composites so that their impact resistance and tolerance can be improved. Some of the methods involve the toughening of the polymer matrix by adding modifiers such as rubber (Bagheri et al. 2009; Yan et al. 2002) and hyper-branched polymers (HPBs) (DeCarli et al. 2005; Verrey et al. 2005). Other methods involve the toughening of the inter-laminar regions of FRP composites using short fiber reinforcements made of Kevlar and Zylon (Sohn et al. 2000; Walker et al. 2002). Through thickness reinforcement methods such as z-pinning (Mouritz 2007) and stitching (Tan et al. 2011, 2012) have also been used to improve the impact performance of FRP composites.

In this chapter, three types of engineered composites and their impact responses are addressed. They are (i) composites with core–shell polymer particles, (ii) composites with carbon nanotubes/nanofibers and (iii) composites with thermoplastic film interleaves.

2   Composites with Core–Shell Polymer Particles

In this section, the effects of core–shell polymer (CSH) particles on the impact resistance of FRP composites are discussed. The CSH particles can be used to toughen the interlaminar regions and improve the energy absorption capabilities of FRP composite laminates. The impact responses of composites with CSH particles were studied by Ali and Joshi (Ali 2014; Ali and Joshi 2012, 2013) and Choo (2014).

Figure 2 shows the schematic diagram of a CSH particle. The inner core of the CSH particle is made of poly butyl acrylate (PBA), which is a soft rubber, whereas the rigid outer shell is made of poly (methyl methacrylate) (PMMA), a thermoplastic. Epoxy functional groups are grafted onto the PMMA outer shell so that the CSH particles can bond well with epoxy resin (Ali 2014).

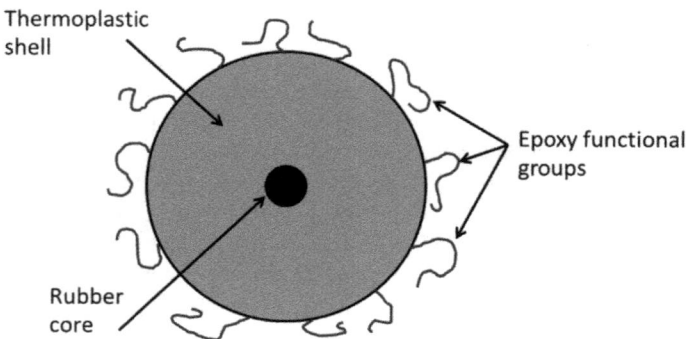

Fig. 2 Structure of the CSH particles

The CSH particles are supplied in white powder form. When observed through a scanning electron microscope (SEM), it can be seen that the CSH particles have varying sizes, ranging from about 20–500 μm. The rubber core has an average size of about 150 nm, while the CSH particle as a whole, including the epoxy functional groups, has an average size of about 160 μm (Ali 2014).

The CSH particles do not agglomerate and are easy to handle compared to other micro- or nano-fillers such as carbon nanotubes (Ali 2014; Ali and Joshi 2012). For the fabrication of FRP composites using the hand layup method, CSH particles can be spread between the FRP prepreg plies before curing. Curing of the FRP composites with CSH particles can be performed in the same manner as composites without CSH particles.

2.1 Impact Response

The impact responses of carbon FRP (CFRP) laminates with and without CSH alteration were investigated by Ali and Joshi (Ali 2014; Ali and Joshi 2012). The test specimens were made using 8 plain weave carbon/epoxy prepregs with fiber volume contents of 60%. For the CSH altered specimens, all of the interlaminar regions (7 in total) were toughened with CSH particles at an areal density of about 50 g/m^2. Drop weight impact tests were carried out at the energy level of 15 J. The impactor used has a hemispherical shape with 12.7 mm diameter. The average velocity at impact was 3.2 m/s, which is in the velocity range for low velocity impact tests. The test specimens with and without CSH alteration were denoted C-CSH50 and C-CSH0 respectively.

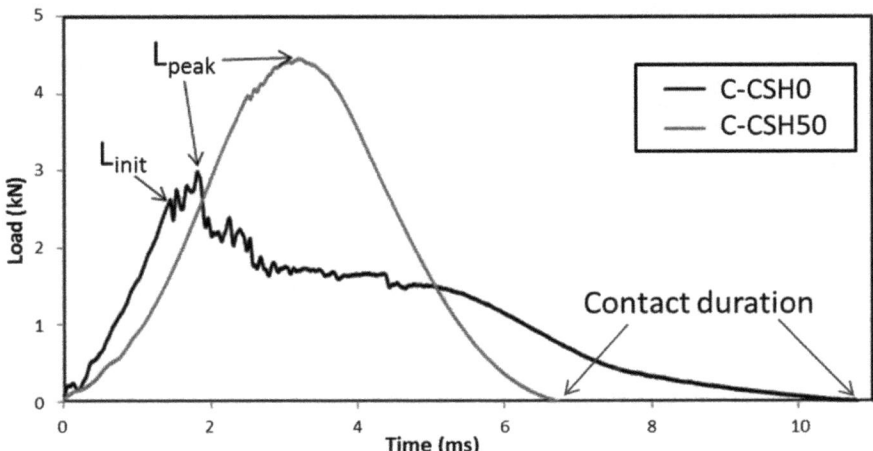

Fig. 3 Impact load responses of C-CSH0 and C-CSH50 laminates

The load-time curves of the impact tests (Ali 2014) are shown in Fig. 3. For both C-CSH0 and C-CSH50 laminates, perforation did not occur. For C-CSH0, damage initiation occurred at the load of 2.52 kN (labeled as L_{init}). After damage initiation, oscillations were recorded in the load-time curve indicating progressive damage taking place in the laminate. The peak load (L_{peak}) occurred at 2.90 kN, followed by a sharp drop in the graph showing that severe damage has occurred (supercritical impact). In contrast, the load-time curve for C-CSH50 has a near symmetrical shape with no significant load drop. This shows that the impact only resulted in minimal damage to the C-CSH50 laminates (subcritical impact). The L_{peak} for C-CSH50 was recorded at 4.09 kN, which is about 41% higher than that of C-CSH50 (Ali 2014; Ali and Joshi 2012).

The extent of damage can also be deduced from the contact duration between the impactor and the test specimens, which can also be determined from Fig. 3. The contact duration for C-CSH0 (average of 10.80 ms) is longer than that of C-CSH50 (average 6.75 ms). The long contact duration for C-CSH0 is due to their larger compliance, which is a consequence of severe impact damage. On the other hand, the C-CSH50 specimens retained high instantaneous stiffness after impact, resulting in short contact duration (Ali 2014; Ali and Joshi 2012).

The energy absorbed during impact for the C-CSH0 and C-CSH50 laminates were also studied (Ali 2014; Ali and Joshi 2012). The addition of CSH particles in the C-CSH50 laminates resulted in an increase of about 230% in E_{elas} and a decrease of about 18% in E_{abs} compared to C-CSH0 laminates. These results show that the CSH particles were very effective in mitigating impact damage in the laminates. The impact test results for the C-CSH0 and C-CSH50 laminates are summarized in Table 1.

Ali and Joshi also studied the performance of CSH altered glass FRP (GFRP) composite laminates subjected to impacts at different energy levels (Ali 2014; Ali and Joshi 2013). Drop weight impact tests (with hemispherical impactor) were carried out at energy levels of about 1.7, 3.3, 4.5, 6.2 and 9.5 J. The test specimens consist of 8 glass/epoxy plies with fiber volume contents of 60% and fibers in the 4-harness satin weave configuration. For the CSH altered specimens (denoted G-CSH30), CSH particles were added to all 7 of the interlaminar regions at the areal density of 30 g/m^2. Reference unaltered specimens (denoted G-CSH0) were also tested for comparison.

Table 1 Summary of impact test results for C-CSH0 and C-CSH50 laminates (Ali 2014)

Characteristics	C-CSH0	C-CSH50
Damage initiation load, L_{init} (kN)	2.52 (9.3)	–
Peak load, L_{peak} (kN)	2.90 (5.5)	4.09 (8.2)
Contact duration (ms)	10.80 (8.9)	6.75 (7.4)
Elastic energy, E_{elas} (J)	1.06 (9.5)	3.48 (5.8)
Absorbed energy, E_{abs} (J)	12.94 (1.4)	10.68 (1.3)

Note Coefficient of variance given in parenthesis

For the G-CSH0 laminates, subcritical response was observed for impact at energy level 1.7 J while higher impact energies resulted in supercritical impact. On the other hand, for G-CSH30 laminates, subcritical response was recorded for impact at energy levels up to 3.3 J. This shows that the CSH particles improved the GFRP laminate's ability to store energy elastically during impact (similar to results for C-CSH50). At the impact energy of 9.5 J, perforation occurred for both G-CSH0 and G-CSH30 laminates (Ali 2014; Ali and Joshi 2013).

Figure 4 shows the peak loads (L_{peak}) recorded during impact at different energy levels for G-CSH0 and G-CSH30 laminates (Ali and Joshi 2013). At low impact energy (1.7 J), L_{peak} values for G-CSH0 and G-CSH30 are about the same because there was very little damage for both cases (subcritical impact). For higher impact energies, CSH particles improved the impact resistance of the laminates, as shown in the higher L_{peak} for G-CSH30. At impact energy 6.2 J and above, the L_{peak} values became almost constant. This means that the ultimate load bearing capacity has been reached. The ultimate load bearing capacity for G-CSH30 is about 60% higher than that of G-CSH0 (Ali 2014; Ali and Joshi 2013).

Ali and Joshi also conducted four point flexural tests on the G-CSH0 and G-CSH30 laminates to study their flexural performances. The flexural modulus and strength of G-CSH30 were 13% and 19% lower than those of G-CSH0 respectively. The addition of CSH particles had a negative effect on the flexural performance of the GFRP laminate (Ali 2014; Ali and Joshi 2013).

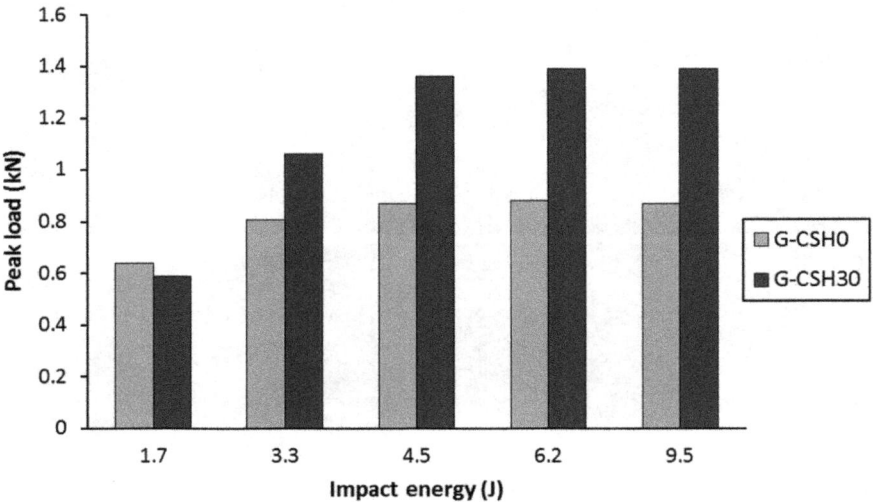

Fig. 4 Peak load versus impact energy for G-CSH0 and G-CSH30 laminates

2.2 Post-impact Analysis

Ali and Joshi studied the interlaminar regions of G-CSH30 laminates after being subjected to impact at 6.2 J using SEM (Ali 2014; Ali and Joshi 2013). The impact energy of 6.2 J resulted in supercritical damage in the G-CSH30 laminates. Figure 5a, b show the cross sections of the damaged laminate taken below the impact point, while Fig. 5c, d show the cross sections at regions away from the impact point. The CSH particles near the point of impact were crushed and broken into smaller pieces. The soft nature of the PMMA shell allows the CSH particles to break under impact load and absorb the impact energy. The broken CSH particles also act to deflect the crack propagation in the epoxy matrix. As a result, cracks in the matrix and delamination are confined to a smaller area of the laminate (Ali 2014; Ali and Joshi 2013). At regions away from the impact point, plastic or tearing deformations in the CSH particles were observed. The tearing deformations consume energy and limit damage to the epoxy matrix due to impact (Ali 2014; Ali and Joshi 2013). These mechanisms combine to improve the impact resistance of the laminate.

Choo conducted CAI tests on GFRP laminates modified with CSH particles to study their impact tolerance (Choo 2014). The laminates studied were made with 8

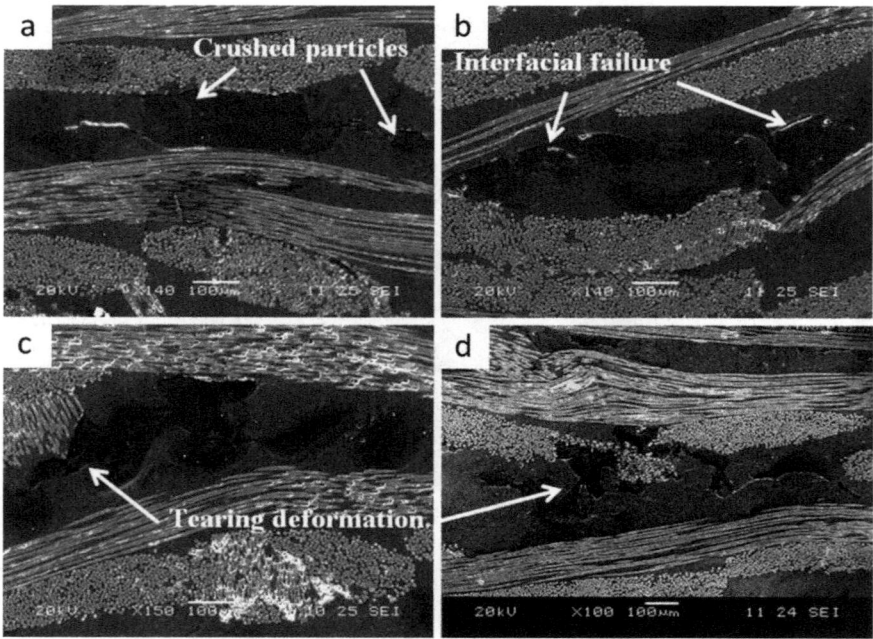

Fig. 5 SEM images of G-CSH30 laminates after supercritical impact: **a** and **b** show cross sectional areas at damaged regions below the point of impact, **c** and **d** show cross sectional areas at regions away from the point of impact (Ali 2014)

layers of glass/epoxy prepregs. CSH particles were added to the interlaminar regions at the areal density of 30 g/m^2. Reference laminates without CSH particles were also fabricated and tested. The laminates with and without CSH particles are designated G-CSH30-C and G-CSH0-C respectively. Drop weight impact tests were carried out at the energy level of 13.4 J. A hemispherical impactor was used. Compressive tests were carried out following the ASTM D3410 standard using a universal testing machine with wedge grips. Compressive test specimens were cut from the impact test specimens (Choo 2014) as drawn in Fig. 6. The compressive test specimens are labeled 1, 2 and 3 based on their positions relative to the impact point (e.g. G-CSH30-C3 is the compressive test specimen furthest away from the impact point, cut from laminates modified with CSH particles).

The compressive moduli and strengths for G-CSH0 and G-CSH30 before and after impact are presented in Table 2 (Choo 2014). In the undamaged state, G-CSH30 had higher compressive strength (increase of about 15%) but lower compressive modulus (decrease of about 30%) compared to G-CSH0. In order to study the impact tolerance of the laminates, the retention factors for the compressive properties were calculated.

Fig. 6 Compressive test specimens cut from impact test specimens

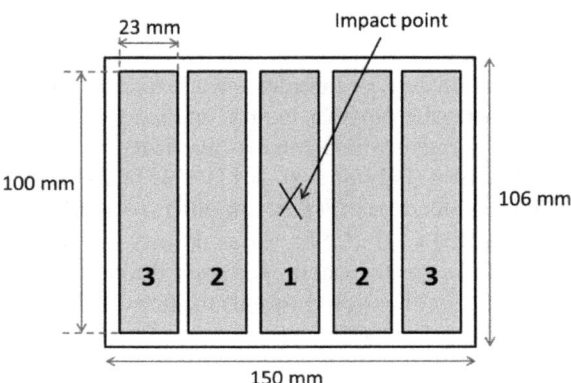

Table 2 Compression-after-impact (CAI) test results for G-CSH0 and G-CSH30 laminates

Specimen	Compressive modulus (MPa)			Compressive strength (MPa)		
	Undamaged (Choo 2014)	Residual (Choo 2014)	Retention factor	Undamaged (Choo 2014)	Residual (Choo 2014)	Retention factor
G-CSH0-C1	710.2	586.3	0.826	207.6	112.8	0.543
G-CSH0-C2	710.2	629.8	0.887	207.6	153.7	0.740
G-CSH0-C3	710.2	667.4	0.940	207.6	208.8	1.01
G-CSH30-C1	494.8	376.9	0.762	239.6	124.1	0.518
G-CSH30-C2	494.8	424.9	0.859	239.6	172.0	0.718
G-CSH30-C3	494.8	464.4	0.939	239.6	239.5	1.00

Note Retention factor = (Residual compressive property)/(Undamaged compressive property)

The retention factors were lower for specimens positioned closer to the point of impact. Overall, the G-CSH30 specimens gave lower retention factors compared to G-CSH0 specimens. This suggests that the CSH particles had a negative effect on the impact damage tolerance of the GFRP laminate.

3   Composites with Carbon Nanotubes/Nanofibers

Carbon-nanotubes (CNTs) and carbon nanofibers (CNFs) are used as nano-fillers for FRP composites because of their high strength and large surface area to volume ratio (Dikshit et al. 2017; Taylor 2010). CNTs and CNFs have been shown to improve the interlaminar fracture toughness of laminated composites (Boon and Joshi 2020; Dikshit et al. 2017; Khan and Kim 2011). Since damage resistance of FRP composites closely relates to the mechanical properties of the interlaminar regions, CNTs and CNFs are also expected to improve the impact performance of FRP composites.

CNTs can be categorized as single walled CNTs (SWCNTs) and multi walled CNTs (MWCNTs) (Taylor 2010). CNTs have a hollow structure and diameters in the order of 1–10 nm. On the other hand, CNFs have a structure made of stacked conical layers and diameters in the order of 100 nm (Kim et al. 2013). CNTs and CNFs can be used in different ways to modify FRP composite laminates, including mixing with the polymer matrix, applied to the interlaminar region (as dry nano-fillers or together with a polymer interleaf), and grafted onto the surface of the fiber reinforcement (Dikshit et al. 2017). One of the main challenges of using CNTs and CNFs is the tendency for the nano-fillers to agglomerate. Without proper dispersion, agglomerated CNTs/CNFs act as defects in FRP composites, leading to negative effects to the mechanical properties of the composites (Taylor 2010).

Researchers have used various methods to disperse and add CNTs/CNFs to FRP composite laminates. One simple method, termed solvent spraying (Fig. 7), is detailed by Dikshit (2014). CNTs are first mixed with ethanol. The mixture is then placed in an ultrasonicator to disperse the CNTs. Subsequently, the mixture is transferred to a sprayer which is used to apply the mixture evenly onto a Teflon sheet. The ethanol solution is allowed to evaporate at elevated temperature, leaving the well dispersed CNTs on the Teflon sheet. The CNTs are then transferred to an FRP composite prepreg by placing the prepreg on top of the Teflon sheet and applying slight pressure using a roller. The prepregs can then be stacked using the hand-layup method and cured to produce composite laminates with CNTs as nano-fillers. The solvent spraying method has been used to achieve good dispersion of CNTs in FRP laminates (Chaudhry et al. 2017; Dikshit 2014; Rodríguez-González et al. 2017). Other methods of dispersing CNTs and CNFs include calendering or the three-roll mill, addition of surfactants, as well as surface functionalization of the nano-fillers (Boon and Joshi 2020).

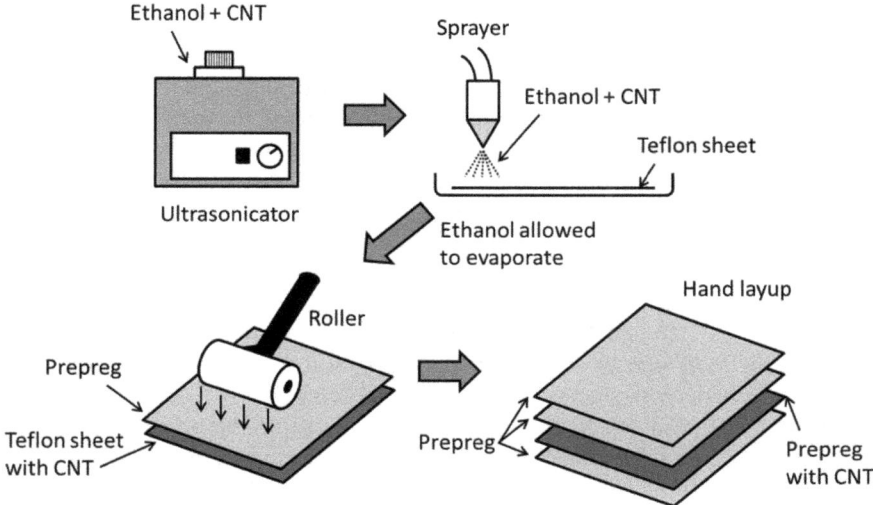

Fig. 7 Schematic of the solvent spraying method

3.1 Impact Response

Kostopoulos et al. studied the impact response of CFRP laminates modified with 0.5 wt.% MWCNT (Kostopoulos et al. 2010). A high shear device was used to disperse the MWCNTs in the epoxy resin. The CFRP laminates have a quasi-isotropic config-uration of $[0/+45/90/-45]_{2s}$ and fiber volume fraction of 58%. Reference laminates with neat epoxy matrix were also tested. Drop weight impact tests were conducted at impact energies 2, 8, 12, 16 and 20 J. The impactor used has a hemispherical shape with 20 mm diameter. For all the impact energies tested, the impactor did not penetrate the specimens. From the impact tests, the peak loads and contact durations for the neat and modified specimens were about the same. At low impact energies of 2 and 8 J, there was no difference between the neat and modified specimens in their energy absorbed and delamination area. At higher impact energies of 16 and 20 J, the modified specimens gave slightly higher energy absorption and slightly smaller delamination area compared to the neat specimens. The authors concluded that CNTs can enhance the impact performance of CFRP laminates at higher impact energy levels.

Rahman et al. studied the impact performance of CFRP laminates modified with 1 wt.% oxidized CNFs (Rahman et al. 2015). The sonication process was used to achieve good dispersion of CNFs in the epoxy resin. The test specimens consist of 16 plain woven layers with fiber volume fraction of about 58%. For comparison, test specimens with neat epoxy matrix were also fabricated. Drop weight impact tests were carried out using a testing machine with a hemispherical impactor (12.5 mm diameter) at impact energies of 10, 20 and 30 J. Rahman et al. found that the peak load during impact for the modified specimens increased by about 11%, 14% and 17% for impact energies of 10, 20 and 30 J respectively compared to the neat specimens. The authors also examined the specimens using ultrasonic C-scan to determine the damage area. The addition of CNFs was found to decrease the damage area due to impact at all three energy levels studied. The largest reduction in damage area (67%) was recorded for the modified specimen subjected to impact at 20 J.

Another study carried out by Taraghi et al. focused on the impact performance of Kevlar/epoxy laminates modified with MWCNT (Taraghi et al. 2014). The MWCNTs were mixed with epoxy resin using a high shear mixer followed by ultrasonication. Woven Kevlar fabrics were used as reinforcements. Laminates with MWCNT contents of 0, 0.3, 0.5 and 1.0 wt.% were prepared by using the hand lay-up technique. Impact testes were carried out at the energy level 45 J using a testing machine with a hemispherical impactor (20 mm diameter). This energy level is higher than the penetration threshold for the unmodified Kevlar/epoxy laminates (30 J). The impact responses of the laminates were studied at room temperature (27 °C) and low temperature (-40 °C) conditions. The impact test results (Taraghi et al. 2014) are plotted in Fig. 8. At room temperature, the specimens with 0.5 wt.% MWCNT gave the highest peak load and absorbed energy, which are about 21% and 35% higher than those of unmodified specimens. However, increasing MWCNT content further resulted in lower peak load and absorbed energy. This is due to agglomeration of MWCNT occurring at high MWCNT content (Taraghi et al. 2014). For low temperature condition, the specimens with 0.3 wt.% MWCNT gave the highest peak load (14% increase) and absorbed energy (35% increase). The epoxy becomes more brittle at lower temperature, thus addition of MWCNT above 0.3 wt.% did not improve the damage resistance further (Taraghi et al. 2014).

3.2 Post-impact Analysis

After impact, Rahman et al. examined the fracture surface of the CFRP laminates modified with CNFs using SEM (Rahman et al. 2015). Fiber bridging due to CNFs were observed for the modified CFRP laminate. This results in larger energy being required to propagate the cracks in the matrix. Sword and sheath fracture of the CNFs were also observed. The fracturing of CNFs absorbs the energy from impact and mitigates damage to the CFRP laminate. Many short and curved cracks were also observed, showing that the CNFs worked to deflect the crack growth. Kostopoulos et al. also used SEM to study the effects of CNTs on the impact response of CFRP laminates (Kostopoulos et al. 2010). They reported that CNT pull-out and fracture occurred in the modified laminates, resulting in larger energy absorbed.

Kosopoulos et al. also performed CAI tests and fatigue compression-after-impact (FCAI) tests on the CFRP laminates modified with 0.5 wt.% MWCNT (Kostopoulos et al. 2010). For CAI tests, the procedure detailed in the ASTM D7137 standard was followed. The addition of MWCNT led to an increase of about 15% in CAI modulus, as well as an increase of 12–15% in CAI strength. The authors attributed

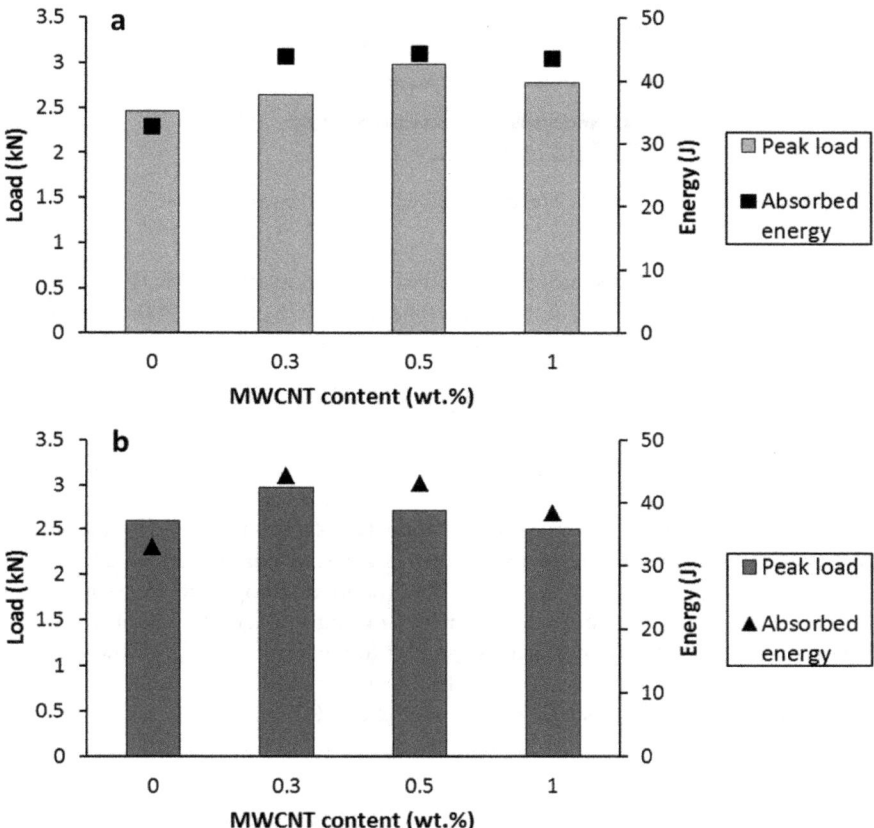

Fig. 8 Impact test (energy level 45 J) results for Kevlar/epoxy laminates with varying MWCNT contents at **a** room temperature (27 °C) and **b** low temperature (−40 °C) conditions

the improved CAI properties to the higher mode I fracture toughness of the MWCNT modified laminates. For the FCAI tests, the MWCNT modified specimens recorded longer fatigue life (by at least 20%) compared to specimens with neat epoxy.

4 Composites with Thermoplastic Film Interleaves

One method to improve the toughness and impact resistance of FRP composites with thermoset matrices is the addition of thermoplastics with high intrinsic toughness. In this section, the toughening of the interlaminar regions in FRP composites using the interleaving method with thermoplastic films is discussed.

Figure 9 shows an FRP laminate toughened using the interleaving method. Researchers have studied the use of many different thermoplastics for interleaving,

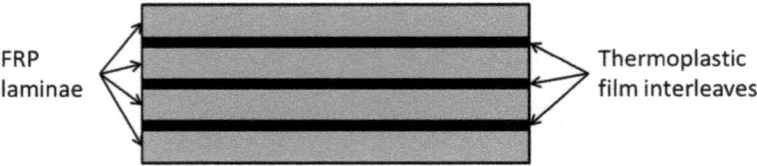

Fig. 9 Schematic of FRP laminate with thermoplastic film interleaves

including ethylene vinyl acetate (EVA) (Eldho 2017), polyetherimide (PEI) (Duarte et al. 1999; Gibson et al. 2001), polyetherketone (PEK) (Cheng et al. 2006), polyethylene (PE) (Tanimoto 1994) and poly(ethylene-co-acrylic acid) (PEAA) (Sohn et al. 2000; Walker et al. 2002).

Depending on the interleaf material, the interleaved FRP laminate can be cured in the same way as non-interleaved laminates (co-curing of interleaf with matrix) (Eldho 2017; Sela and Ishai 1989) or with an extra post-curing step (Tanimoto 1994). After curing, the thermoplastic interleaves remain as discrete layers in the laminate (Sela and Ishai 1989). The fabrication process for interleaved laminates remain simple because the properties of the polymer matrix (such as viscosity) remain unchanged during processing (Cheng et al. 2006). The tough thermoplastic interleaves can undergo plastic deformation to absorb energy during impact (Walker et al. 2002). However, the thermoplastic interleaves used for toughening often have low stiffness and strength, leading to lower specific stiffness and strength for the interleaved FRP laminate as well (Sela and Ishai 1989; Walker et al. 2002).

4.1 Impact Response

Duarte et al. studied the impact responses of carbon/epoxy laminates with different types of thermoplastic interleaves (Duarte et al. 1999). The researchers fabricated interleaved CFRP laminates with lay-up $[+45/I/0/I/-45/I/90/I/+45/I/0/I/-45/I/90]_S$ where I refers to the thermoplastic interleaves. The types of interleaves investigated are detailed in Table 3. Non-interleaved CFRP laminates with lay-up $[+45/0/-45/90]_{2S}$ were also fabricated and tested for comparison. Drop weight impact tests were carried out at energy levels of about 1, 2.5, 4 and 6 J/mm. The impactor used was hemispherical with a diameter of 12.5 mm.

Duarte et al. measured the damage area of the impact test specimens using C-scan (Duarte et al. 1999). The specimens with XAF2210 interleaves resulted in smaller damage area compared to the non-interleaved specimens for all impact energy levels studied. In particular, at 6 J/mm, the damage area for XAF2210 specimens was about 55% smaller than that of the non-interleaved specimens. The specimens with the perforated interleaves XAF2210P also resulted in improved impact resistance but the damage area was larger compared to the XAF2210 specimens. On the other hand, the XAF2065 interleaves only gave slight improvement to the impact resistance

Table 3 Thermoplastic interleaves studied by Duarte et al. (1999)

Designation	Material	Molded thickness (mm)	Melting temperature (°C)	Interleaf structure
1a8s18	Copolyamide	0.018	145	Web-like structure
XAF2210	Polyolefin	0.030	145	Thin film
XAF2210P	Polyolefin	0.030	145	Perforated XAF2210, hole size 0.25 in, at 20 mm staggered spacing
XAF2065	Polyolefin	0.039	125	Open net form of XAF2210
ULTEM 1000	Polyetherimide	0.048	225	Thin Film

of the CFRP laminates. For instance, at impact energy of 6 J/mm, the XAF2065 specimens resulted in only about 15% smaller damage area compared to that of the non-interleaved specimens. Therefore, among the specimens with polyolefin interleaves, the specimens with the film type interleaves (XAF2210) showed the best impact resistance.

For the specimens with 1a8s18 interleaves, Duarte et al. reported that the damage area was larger than that of the non-interleaved specimens (Duarte et al. 1999). This was attributed to the web-like structure of the 1a8s18 interleaves and the poor bonding between the copolyamide and the epoxy matrix. The poor bonding resulted in the 1a8s18 interleaves acting as weak points in the laminate for crack initiation. On the other hand, improved impact resistance was achieved with the polyetherimide interleaves (ULTEM 1000). Impact damage was observed on the specimens with ULTEM 1000 interleaves starting at the impact energy of 2.7 J/mm. For impact energies between 2.7 and 5 J/mm, the damage area for the specimens with ULTEM 1000 interleaves was smaller than that of non-interleaved specimens. However, for impact energies above 5 J/mm, the impact resistance of the laminates was not affected by the ULTEM 1000 interleaves. The authors explained that this is due to fiber breakage being the main damage mode for laminates subjected to impact at high energies (Duarte et al. 1999).

Another study on the impact response of CFRP laminates toughened with thermoplastic interleaves was performed by Cheng et al. (2006). The composite laminates studied were made of carbon fiber as reinforcement and bismaleimide (BMI) resin as matrix. Polyetherketone with a phenolphtalein side group (denoted PEK-C) was used as the toughening agent. Cheng et al. studied 4 types of CFRP laminates where the matrix components were toughened with different methods. The 4 types of specimens are (i) specimens with neat BMI as matrix (control case), (ii) specimens with BMI/PEK-C blend as matrix (termed in-situ toughening), (iii) specimens with BMI matrix and PEK-C film interleaves, and (iv) specimens with BMI matrix and BMI/PEK-C blend film interleaves. For the PEK-C toughened specimens, the PEK-C content in the specimens was set to 17.5 wt.%. The specimens were subjected to

Table 4 Impact experiment results of carbon/BMI laminates with different thermoplastic toughening methods studied by Cheng et al. 2006

Case	Composition of matrix	Damage area (mm^2) Cheng et al. (2006)
TP0	BMI matrix (control case)	544
TP1	BMI/PEK-C blend as matrix, PEK-C content at 17.5 wt.%	408
TP2	BMI matrix with PEK-C film interleaves, overall PEK-C content at 17.5 wt.%	345
TP3	BMI matrix with BMI/PEK-C blend film interleaves, overall PEK-C content at 17.5 wt.%	220

impact at impact energy 2 J/mm. The delamination area due to impact was determined using ultrasonic C-scan. The impact test results are summarized in Table 4. Specimens toughened by film interleaving were found to perform better than specimens with in-situ toughening. For film interleaving, the BMI/PEK-C blend film was reported to give smaller impact damage area than the PEK-C film.

4.2 Post-impact Analysis

Duarte et al. studied the undamaged compression strengths and CAI strengths of the CFRP laminates with different thermoplastic interleaves (Table 3) (Duarte et al. 1999). For specimens with XAF2210 interleaves, the undamaged compression strength was 65% lower than that of non-interleaved specimens. Further investigation using a microscope revealed that fiber micro buckling occurred in the specimens leading to the low compression strength. This was due to the low shear modulus of the polyolefin interleaf. The CAI performance of the specimens with XAF2210 interleaves was also worse than non-interleaved specimens. Therefore, although the XAF2210 interleaves reduced the impact damage area, their low shear modulus resulted in significant reduction to the compression strength of the CFRP laminate.

In contrast, CFRP laminates with ULTEM 1000 interleaves had higher undamaged compression strength compared to non-interleaved laminates. This is because of the high shear modulus of polyetherimide (Duarte et al. 1999). Microscopic studies of the ULTEM 1000 interleaved specimens after impact revealed that cracks tend to propagate in the transverse direction through the lamina but not along the interlaminar regions. The ULTEM 1000 interleaves resulted in higher interlaminar fracture toughness for the interleaved specimens. For CAI tests, at impact energy of 1 J/mm, the ULTEM 1000 interleaved specimens gave very high residual strength (93% of undamaged strength) compared to non-interleaved specimens (70% of undamaged strength). However, the improvement in CAI strength became less significant at higher impact energy levels. For instance, ULTEM 1000 interleaved specimens gave 53% residual strength after 4.1 J/mm impact whereas non-interleaved specimens resulted in 40% residual strength after 4.3 J/mm impact (Duarte et al. 1999).

Table 5 Compression-after-impact (CAI) test results for carbon/BMI laminates toughened with PEK-C

Case	Composition of matrix	CAI strength (MPa) Cheng et al. (2006)
TP0	BMI (control case)	180
TP1	BMI/PEK-C blend, PEK-C content at 17.5 wt.%	199
TP2	BMI matrix with PEK-C film interleaves, overall PEK-C content at 17.5 wt.%	254
TP3-a	BMI matrix with BMI/PEK-C blend film interleaves, interleaf weight ratio of BMI:PEK-C = 10:90	256
TP3-b	BMI matrix with BMI/PEK-C blend film interleaves, interleaf weight ratio of BMI:PEK-C = 20:80	272
TP3-c	BMI matrix with BMI/PEK-C blend film interleaves, interleaf weight ratio of BMI:PEK-C = 40:60, overall PEK-C content at 17.5 wt.%	290

Cheng et al. also studied the impact tolerance of carbon/BMI laminates toughened with PEK-C using CAI tests with impact energy 2 J/mm (Cheng et al. 2006). The CAI test results are summarized in Table 5. For specimens toughened with BMI/PEK-C blend film (cases TP3-a, TP3-b and TP3-c in Table 5), the researchers studied three different blend compositions for the film interleaves. The specimens with BMI/PEK-C blend film interleaves at weight ratio of BMI: PEK-C = 40:60 (case TP3-c) gave the highest CAI strength of 290 MPa. The morphology of the interlaminar regions for the carbon/BMI laminates toughened with BMI/PEK-C blend film interleaves was studied using SEM. The researchers reported a fine granular morphology (average diameter of 1 μm) at the interlaminar regions of the laminates (Cheng et al. 2006). The granular morphology was not present for the laminates toughened using PEK-C film interleaves (case TP2). This suggests that the morphology at the interlaminar regions of composite laminates is important in determining their impact resistance and tolerance.

5 Conclusions

CSH particles have been shown to be effective in improving the impact resistance of FRP composite laminates. In particular, laminates modified with the particles showed improved elastic energy storing capabilities during impact. The soft outer shell of the CSH particles generally gets crushed during impact, absorbing energy in the process. Other strengthening mechanisms due to the particles include crack deflection and plastic deformation. These mechanisms combine to hinder crack propagation and limit the severity of impact damage sustained by the modified laminates. However, impact damage tolerance is not improved by the addition of CSH particles.

Researchers have modified FRP composites using CNTs/CNFs and achieved varying levels of improvement to the impact resistance and tolerance of the composites. Fiber bridging and pull-out due to the nano-fillers absorb large amounts of energy, thus reducing the energy available for damage creation in the modified laminates. The nano-fillers also work to hinder crack propagation through crack deflection. Studies have shown that CNTs/CNFs are effective at improving the impact resistance of FRP composite laminates subjected to impact at high energy levels.

The interleaving method is a simple way to toughen the interlaminar regions of FRP composites. For interleaving with thermoplastic films, the improvement to the impact performance of the FRP composite is affected by various factors, including the mechanical properties of the thermoplastic and the resulting morphology of the interlaminar regions. Besides having high toughness, the thermoplastic also needs to have high shear modulus for the thermoplastic film interleave to be effective in toughening the composites.

Comparing the three types of engineered composites, the composites with CSH particles and composites with thermoplastic film interleaves are preferred when ease of manufacturing is important. However, CSH particles and thermoplastic film interleaves both result in reduced in-plane mechanical properties in the composite. This reduction needs to be compensated for with thicker or larger structures, leading to increased weight. For composites with CNTs/CNFs, further studies need to be carried out to optimize their impact performances. In particular, the dispersion process of nano-fillers needs to be further improved to enable higher nano-filler content in the modified composite laminate.

References

Abrate S (1991) Impact on laminated composite materials. Appl Mech Rev 44:155–190. https://doi.org/10.1115/1.3119500

Ali M (2014) Impact behaviour of prepreg laminates with dispersed core/shell particles in ply interfaces. Nanyang Technological University, Singapore

Ali M, Joshi SC (2012) Impact damage resistance of CFRP prepreg laminates with dispersed CSP particles into ply interfaces. Int J Damage Mech 21:1106–1127. https://doi.org/10.1177/1056789511429143

Ali M, Joshi SC (2013) Damage evolution in glass/epoxy composites engineered using core–shell microparticles under impact loading Journal of Materials Science 48:8354–8367 doi:https://doi.org/10.1007/s10853-013-7635-8

Bagheri R, Marouf BT, Pearson RA (2009) Rubber-toughened epoxies: a critical review. Polym Rev 49:201–225. https://doi.org/10.1080/15583720903048227

Boon YD, Joshi SC (2020) A review of methods for improving interlaminar interfaces and fracture toughness of laminated composites Materials Today. Communications 22:100830. https://doi.org/10.1016/j.mtcomm.2019.100830

Cantwell WJ, Morton J (1991) The impact resistance of composite materials — a review Composites 22:347–362 doi:https://doi.org/https://doi.org/10.1016/0010-4361(91)90549-V

Chaudhry MS, Czekanski A, Zhu ZH (2017) Characterization of carbon nanotube enhanced interlaminar fracture toughness of woven carbon fiber reinforced polymer composites. Int J Mech Sci 131–132:480–489. https://doi.org/10.1016/j.ijmecsci.2017.06.016

Cheng Q, Fang Z, Xu Y, Yi X-S (2006) Improvement of the Impact Damage Resistance of BMI/Graphite Laminates by the Ex-situ Method. High Perform Polym 18:907–917. https://doi.org/10.1177/0954008306068296

Choo KCJ (2014) Compression after impact studies on composite laminates. Nanyang Technological University, Singapore

DeCarli M, Kozielski K, Tian W, Varley R (2005) Toughening of a carbon fibre reinforced epoxy anhydride composite using an epoxy terminated hyperbranched modifier. Compos Sci Technol 65:2156–2166. https://doi.org/10.1016/j.compscitech.2005.05.003

Dikshit V (2014) Manufacturing and performance studies of laminated composites with nano-reinforced inter-ply interfaces. Nanyang Technological University, Singapore

Dikshit V, Bhudolia S, Joshi S (2017) Multiscale Polymer Composites: A Review of the Interlaminar Fracture Toughness Improvement Fibers 5:38

Duarte A, Herszberg I, Paton R (1999) Impact resistance and tolerance of interleaved tape laminates. Compos Struct 47:753–758. https://doi.org/10.1016/S0263-8223(00)00049-0

Eldho M (2017) Effects of thermoplastic film interleaving on fracture toughness of GFRP laminates. Nanyang Technological University, Singapore

Gibson RF, Chen Y, Zhao H (2001) Improvement of Vibration Damping Capacity and Fracture Toughness in Composite Laminates by the Use of Polymeric Interleaves. J Eng Mater Technol 123:309–314. https://doi.org/10.1115/1.1370385

He W, Lu S, Yi K, Wang S, Sun G, Hu Z (2019) Residual flexural properties of CFRP sandwich structures with aluminum honeycomb cores after low-velocity impact. Int J Mech Sci 161–162:105026. https://doi.org/10.1016/j.ijmecsci.2019.105026

Khan SU, Kim J-K (2011) Impact and delamination failure of multiscale carbon nanotube-fiber reinforced polymer composites: a review international Journal of Aeronautical and Space. Science 12:115–133. https://doi.org/10.5139/IJASS.2011.12.2.115

Kim YA, Hayashi T, Endo M, Dresselhaus MS (2013) Carbon nanofibers. In: Vajtai R (ed) Springer Handbook of Nanomaterials. Springer, Berlin, Heidelberg, pp 233–262

Kostopoulos V, Baltopoulos A, Karapappas P, Vavouliotis A, Paipetis A (2010) Impact and after-impact properties of carbon fibre reinforced composites enhanced with multi-wall carbon nanotubes. Compos Sci Technol 70:553–563. https://doi.org/10.1016/j.compscitech.2009.11.023

Kumar P, Rai B (1993) Delaminations of barely visible impact damage in CFRP laminates. Compos Struct 23:313–318. https://doi.org/10.1016/0263-8223(93)90231-E

Mouritz AP (2007) Review of z-pinned composite laminates. Compos A Appl Sci Manuf 38:2383–2397. https://doi.org/10.1016/j.compositesa.2007.08.016

Rahman MM, Hosur M, Hsiao K-T, Wallace L, Jeelani S (2015) Low velocity impact properties of carbon nanofibers integrated carbon fiber/epoxy hybrid composites manufactured by OOA–VBO process Composite Structures 120:32–40. https://doi.org/10.1016/j.compstruct.2014.09.053

Rodríguez-González JA, Rubio-González C, Meneses-Nochebuena CA, González-García P, Licea-Jiménez L (2017) Enhanced interlaminar fracture toughness of unidirectional carbon fiber/epoxy composites modified with sprayed multi-walled carbon nanotubes. Compos Interfaces 24:883–896. https://doi.org/10.1080/09276440.2017.1302279

Santiuste C, Sanchez-Saez S, Barbero E (2010) Residual flexural strength after low-velocity impact in glass/polyester composite beams. Compos Struct 92:25–30. https://doi.org/10.1016/j.compstruct.2009.06.007

Sela N, Ishai O (1989) Interlaminar fracture toughness and toughening of laminated composite materials: a review. Composites 20:423–435. https://doi.org/10.1016/0010-4361(89)90211-5

Sohn MS, Hu XZ, Kim JK, Walker L (2000) Impact damage characterisation of carbon fibre/epoxy composites with multi-layer reinforcement. Compos B Eng 31:681–691. https://doi.org/10.1016/S1359-8368(00)00028-7

Tan KT, Watanabe N, Iwahori Y (2011) Impact damage resistance, response, and mechanisms of laminated composites reinforced by through-thickness stitching. Int J Damage Mech 21:51–80. https://doi.org/10.1177/1056789510397070

Tan KT, Watanabe N, Iwahori Y, Ishikawa T (2012) Understanding effectiveness of stitching in suppression of impact damage: an empirical delamination reduction trend for stitched composites. Compos A Appl Sci Manuf 43:823–832. https://doi.org/10.1016/j.compositesa.2011.12.022

Tanimoto T (1994) Suppression of interlaminar damage in carbon/epoxy laminates by use of interleaf layers. Scr Metall Mater 31:1073–1078. https://doi.org/10.1016/0956-716X(94)90529-0

Taraghi I, Fereidoon A, Taheri-Behrooz F (2014) Low-velocity impact response of woven Kevlar/epoxy laminated composites reinforced with multi-walled carbon nanotubes at ambient and low temperatures. Mater Des 53:152–158. https://doi.org/10.1016/j.matdes.2013.06.051

Taylor AC (2010) Advances in nanoparticle reinforcement in structural adhesives. In: Dillard DA (ed) Advances in structural adhesive bonding. Woodhead Publishing Limited, Cambridge, UK, pp 151–182

Verrey J, Winkler Y, Michaud V, Månson JAE (2005) Interlaminar fracture toughness improvement in composites with hyperbranched polymer modified resin. Compos Sci Technol 65:1527–1536. https://doi.org/10.1016/j.compscitech.2005.01.005

Walker L, Sohn M-S, Hu X-Z (2002) Improving impact resistance of carbon-fibre composites through interlaminar reinforcement. Compos A Appl Sci Manuf 33:893–902. https://doi.org/10.1016/S1359-835X(02)00010-6

Yan C, Xiao K, Ye L, Mai Y-W (2002) Numerical and experimental studies on the fracture behavior of rubber-toughened epoxy in bulk specimen and laminated composites. J Mater Sci 37:921–927. https://doi.org/10.1023/A:1014335511515

The Potential of Biocomposites in Low Velocity Impact Resistance Applications

Fabrizio Sarasini, Jacopo Tirillò, Claudia Sergi, and Francesca Sbardella

Abstract The urgent need to make natural fibre composites suitable for semi-structural applications demands a thorough assessment of their behaviour under different loading conditions and strain rates. In this regard, low velocity impact represents a severe hazard to the composite industry due to the resulting complex damage scenario able to markedly impair the mechanical properties of composite structures. The aim of this chapter is to provide a comprehensive review of the resistance to low velocity impacts of natural fibre composites, with a view to highlighting the effects of the various factors that influence the impact resistance of traditional fibre reinforced composites. The potential of natural fibre composites and differences with the behaviour of the synthetic counterparts are addressed, along with the areas that need improvement for a better exploitation of natural fibre composites in semi- or structural applications. Literature survey highlighted that also for natural fibre composites the toughness of the matrix dictates the energy absorbed at perforation, the damage resistance and tolerance, which are largely independent of fibre architecture. Another important feature, for energies far from perforation, is the less detrimental role played by delamination compared to synthetic laminates.

Keywords Natural fibres · Natural fibre composites · Low velocity impact · Impact resistance · Damage tolerance

1 Introduction

Over the last twenty years the use of biocomposites, intended as conventional polymer matrices sourced from fossil resources reinforced with natural fibres, has recorded an enduring increase in several industries due to global awareness and promotion of sustainable development (Pickering et al. 2016; Sanjay et al. 2018; Gholampour and Ozbakkaloglu 2020). In this regard, the global natural fibre composites market size was esteemed at USD 4.46 billion in 2016, while it is envisaged to grow with a

F. Sarasini (✉) · J. Tirillò · C. Sergi · F. Sbardella
Department of Chemical Engineering Materials Environment, Sapienza-Università Di Roma, Via Eudossiana 18, 00184 Roma, Italy
e-mail: fabrizio.sarasini@uniroma1.it

© Springer Nature Singapore Pte Ltd. 2021
M. T. H. Sultan et al. (eds.), *Impact Studies of Composite Materials*, Composites Science and Technology, https://doi.org/10.1007/978-981-16-1323-4_8

CAGR (Compound Annual Growth Rate) of 11.8% from 2016 to 2024, in accordance with a recent market report (Grand View Research Inc. 2018). The most important raw materials are wood, flax, kenaf, cotton, and hemp, but wood is still governing the market with a 59.3% share of the total revenue in 2015 (Grand View Research Inc. 2018). Another important contribution is offered by flax, which in 2015 had a market share of 13.0% thanks to its CO_2 neutrality, vibration damping ability and high specific mechanical properties (Yan et al. 2014; Bourmaud et al. 2018).

The need for lightweight and sustainable composite materials is governing the rise of applications of natural fibre composites in two specific industrial sectors, namely the automotive and construction, where biocomposites are usually applied in cosmetic applications (door panels, decking, frames, etc.). Natural fibre composites find applications in the automotive field because the parts, in addition to adequate mechanical performance, offer a weight saving (by 30%) that allows to reduce the fuel consumption and diminish CO_2 emissions. It is not surprising that this segment accounted for a revenue share of over 30% in 2015 (Grand View Research Inc. 2018), the main objective being the replacement of glass fibres with wood or non-wood fibres such as flax and hemp (Koronis et al. 2013). Natural fibres represent also inexpensive and sustainable alternatives to synthetic fibres used as building materials, supported by a share of 56% of the overall market volume in 2015 (Dittenber and GangaRao 2012; Grand View Research Inc. 2018).

This development, especially in the transportation field, has exacerbated the need for a thorough understanding of the damage and fracture mechanisms of natural fibre composites when subjected not only to quasi-static but also to dynamic loads. The shift from static to dynamic loading in heterogeneous and anisotropic materials, like composites, is much more complicated compared to traditional metallic materials. Two or more constituents with varying mechanical properties and potentially different fibre/matrix adhesion quality influence the propagation of stress waves that in turn results in a complex and unpredictable damage initiation and subsequent propagation. In this regard, a major threat to composite structures is represented by their proneness to low velocity impacts, as they produce significant internal damage as delaminations, matrix cracks and fibre breakages, which can go easily undetected but that considerably influence their residual mechanical properties. This occurs also if barely visible impact damage (BVID) is produced. In fact, BVID can involve delaminations and back-face splitting, which result in residual strength reductions by as much as 50–60% (Shah et al. 2019). It is therefore of utmost importance to prove that composite structures can bear loads even when already damaged, which is included in the well-known issue of damage tolerance. The problem with composite materials lies in the meaning of the word damage, as it involves an accumulation of matrix cracks of different size-scales, shapes and orientations, voids, broken fibres and delaminations. Therefore it is expected that this "state of damage" can originate the loss of diverse design-induced functionalities and the loss of strength is just one of them (Talreja and Phan 2019). This scenario is even much more complex for natural fibre composites, in which the inherent variability in natural fibre properties is often coupled with a non-reliable and in-depth understanding of their mechanical

behaviour, in particular in terms of their impact damage resistance and damage tolerance. This complexity can partly explain why information on low velocity impact
behaviour of composites including natural fibres, such as jute, flax, hemp, etc., is still
quite restricted. Nevertheless, this is a valuable and required property for assessing
their suitability to semi- or structural applications. The aim of this chapter is to
outline a comprehensive overview of the low velocity impact behaviour of natural
fibre composites with reference to the several factors that influence the impact resistance of traditional fibre reinforced composites (Fig. 1), to identify the missing points

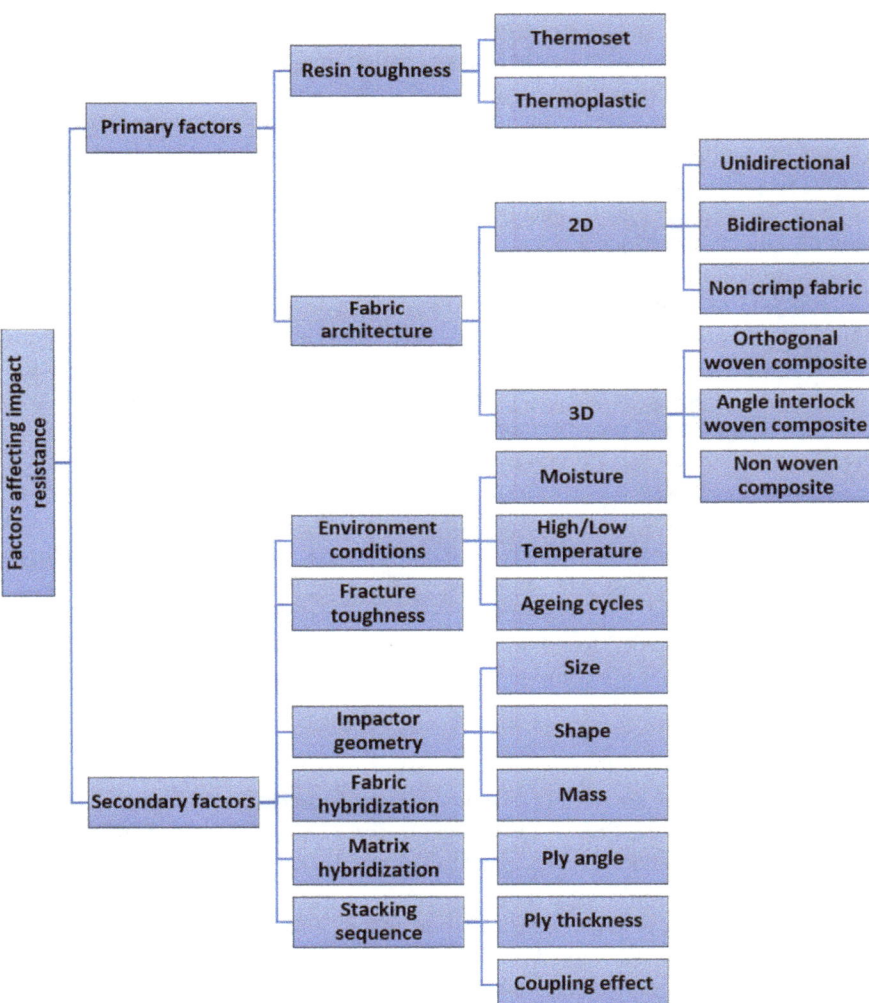

Fig. 1 Factors affecting impact resistance of fibre reinforced composites (adapted from Shah et al.
(2019))

in the available literature and recommend potential directions for future researches in this field.

2 Low Velocity Impact of Natural Fibre Composites

2.1 Effect of Matrix Toughness on Impact Behaviour

Among the different factors summarized in Fig. 1, matrix toughness represents an important one, and its effect has been investigated in conventional laminates based on synthetic fibres. From the available results, composites based on thermoplastic matrices usually show a better impact resistance compared to thermoset-based ones in terms of smaller delaminated areas (Jang et al. 1991; Schrauwen and Peijs 2002; Vieille et al. 2013; Arikan and Sayman 2015). Matrix plasticization is particularly active in matrix-rich zones because it stimulates local deformation, while fibre-bridging hinders the Mode I opening of plies and retards the growth of interlaminar and intralaminar cracks according to Mode II and Mode III. The overall result is a reduction of their global growth that is accompanied with the development of limited delaminations in size (Vieille et al. 2013). This is the typical damage scenario created by an impact: at first the high out-of-plane shear stresses under the impactor-surface contact point generate matrix cracks that once they reach the lower interface, promote the opening of the interface and trigger the damage following the Mode I. The delamination grows mainly under Mode II.

This trend seems to be confirmed also in natural fibre composites. Bensadoun et al. (2017a) performed a detailed investigation on flax fibre reinforced composites with a view to highlighting the effect of matrix type (epoxy and polypropylene (PP)) on impact properties. The authors used different levels of impact energy up to perforation, and found that for both matrices a traditional power law can be used to foretell the energy needed to achieve perforation, the same that has been validated with glass and carbon fibre composites (Eq. 1) (Caprino and Lopresto 2001):

$$U_{perforation} = K\left(t V_f D_t\right)^\alpha \tag{1}$$

where K and α are two material constants to be experimentally determined, t is the thickness in mm, V_f the fibre volume fraction and D_t the diameter of the striker in mm. The calculated value of α was equal to 1.3, exactly the one reported for synthetic composites (Caprino and Lopresto 2001).

The type of matrix played a significant role, as confirmed by the value of parameter K, which was equal to 12.5 and 9 for flax-thermoplastic (PP) and for flax-thermoset (epoxy) composites, respectively, thus suggesting a better behaviour of thermoplastic composites and the possibility to accurately predict the perforation energy once the thickness and the fibre volume fraction are known. During an impact that causes perforation, there is the matrix and fibres breakage, and therefore also their role has to be considered. In the specific case of natural fibre composites, natural fibres are not as strong as glass or carbon ones, therefore their contribution to the energy needed to perforate a specimen is lower than the matrix toughness. The high ductility of PP increased the perforation energy, an increase dependent on the fibre architecture, in the range 20–50% compared to thermoset based composites. In thermoplastic composites, most of the energy is dissipated after the peak force, suggesting a higher resistance experienced by the impactor while penetrating the specimens.

In another study, where an epoxy matrix reinforced with hemp fibres was compared with a polylactic acid (PLA) matrix (Caprino et al. 2015), the power law (Eq. 1) was not found to hold. In this case, a linear trend was detected (Eq. 2):

$$U_{perforation} = a \cdot \left(t V_f D_t \right) + b \tag{2}$$

where the a and b constants were reported to be equal to 0.5 and 13.4, respectively.

Interestingly, when correlating the absorbed energy (U_a) with the impact energy (U), the authors found a very simple linear relationship for both epoxy and PLA matrices (Eq. 3):

$$U_a = a \cdot U - b \tag{3}$$

The values of the material parameters, a and b, were reported as 0.745 and 3.66 for epoxy-based composites, and 0.653 and 2.71 for PLA-based composites, respectively. A similar relationship was obtained in (Sutherland and Guedes Soares 2005) for glass/polyester composites. When $U_a = 0$, the threshold impact energy (U_0) below which no energy is absorbed to produce damage can be estimated. This parameter was equal to 4.9 and 4.1 J for epoxy and PLA-based composites, respectively, higher than corresponding values for glass fibre reinforced composites (Caprino et al. 2011), thus implying that higher energies are potentially required in natural fibre composites for the onset of damage compared to synthetic laminates.

In both studies (Caprino et al. 2015; Bensadoun et al. 2017a), in non-perforating impacts, the thermoset systems exhibited greater damage area compared to thermoplastic composites. A severe through-the-thickness crack with fibre failure and limited delaminations were detected in thermoset and thermoplastic composites (Fig. 2), mainly related to the inherent low strength of the flax fibres and to the high interlaminar fracture toughness (G_{Ic}) of the flax composites fostered by specific energy absorbing mechanisms, such as crack branching and fibre bridging (Bensadoun et al. 2017b). The brittle behaviour of epoxy matrix exacerbated the presence of matrix cracks compared to delaminations, while in thermoplastic composites a more concentrated impact damage zone was detected, and the higher ductility allowed a higher energy absorption and restrained the development of cross-shaped cracks in the composites.

This behaviour represents a significant difference with respect to synthetic fibre composites (Liu and Hughes 2008). In the tensile side there is the origin of cracks oriented perpendicularly to the direction of fibres that later grow through the thickness and result in small delaminations. In conventional laminates, the glass or carbon

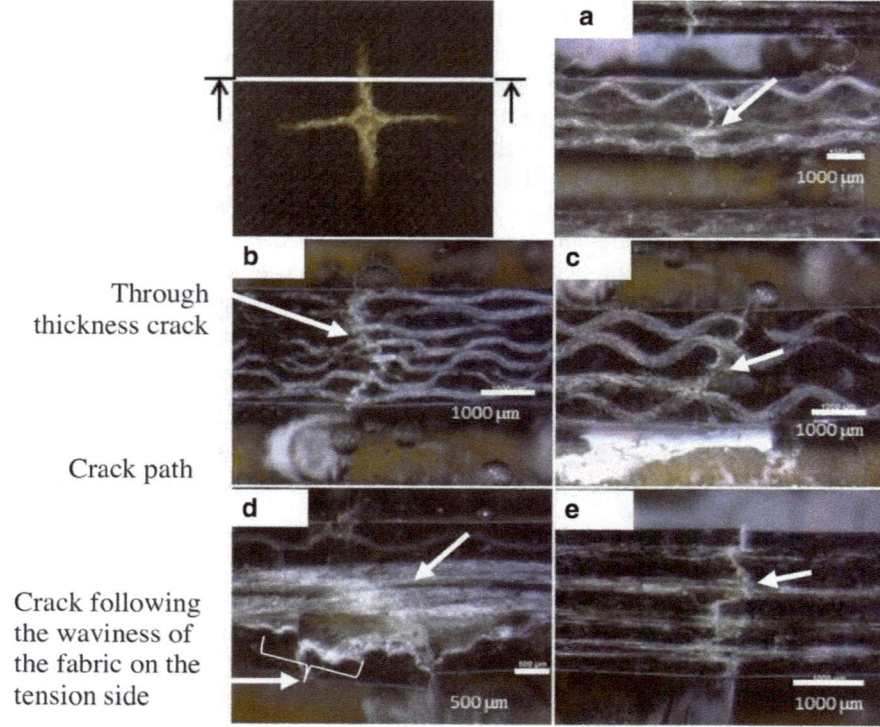

Through thickness crack

Crack path

Crack following the waviness of the fabric on the tension side

Fig. 2 Through-the-thickness damage in flax/epoxy and flax/PP composites after a 3.1 J impact: **a** plain weave with PP; **b** medium–high twist twill with epoxy; **c** plain weave with epoxy; **d** quasi-UD [0, 90] with epoxy; **e** UD [0, 90] with epoxy (reprinted with permission from Bensadoun et al. (2017a))

fibres are less prone to be fractured and only matrix cracks are generated, which can subsequently start delaminations (Shyr and Pan 2003).

To assess the impact damage tolerance, the authors (Bensadoun et al. 2017a) did not perform the standard compression after impact (CAI) tests, because of extensive buckling of the specimens. In particular, they used a flexure after impact approach, well used in literature for synthetic laminates (Sarasini et al. 2014). The reduced damage in thermoplastic-based composites resulted in a marginal decrease in flexural properties after impact and in any case lower compared to that experienced by thermoset-based composites.

2.2 Effect of Fabric Architecture and Stacking Sequence on Impact Behaviour

Fabric architecture is another key parameter influencing the impact resistance and damage tolerance of composite laminates. Unidirectional laminates (UD) offer superior quasi-static mechanical properties but suffer from poor impact resistance (Shah et al. 2019), and 2D laminates represent a better alternative due to the yarn waviness in the fabric architecture. Unfortunately, their in-plane mechanical properties are significantly lower than UDs because the crimps in the yarns act as stress concentration points. This explains the development of non-crimp fabrics, which brought improvements in delamination resistance (Greve and Pickett 2006). Delamination that can be further hindered by using techniques that introduce reinforcement through-the-thickness, such as stitching, z-pinning, etc. (Mouritz 2001, 2007; Francesconi and Aymerich 2017; Yasaee et al. 2017), at the expense of in-plane mechanical properties.

3D woven composites represent a relatively recent development in structures subjected to impact, because the yarns that link together the different plies hinder the development of large delaminations. Few experimental studies have addressed influence of such architecture on impact and post-impact behaviour of composites (Bibo and Hogg 1996; Chiu et al. 2004; Chen and Hodgkinson 2009; Seltzer et al. 2013; Elias et al. 2017). Seltzer et al. (2013) showed that 3D composites dissipated over twice the energy compared to 2D laminates and that this energy absorption capability was essentially affected by the z-yarns, which introduced energy dissipation mechanisms such as tow splitting, fibre breakage under the tup and creation of a plug by out-of-plane shear.

Contrary to synthetic fibre reinforced composites, the effects of stitching on the mechanical properties of natural fibre composites have not been widely investigated. Rong et al. (2002) analysed the factors affecting in-plane mechanical response and Mode I interlaminar fracture toughness of laminates based on an epoxy matrix reinforced with unidirectional sisal fibres but stitched with Nylon 6,6, Kevlar and sisal threads. The in-plane mechanical properties of sisal laminates were not degraded by the presence of threads because the sisal fibres showed a high damage tolerance, but at the same time the presence of stitches expanded the fibre bridging zone and improved the resistance to delamination.

Ravandi et al. (2016) addressed the effect of stitch areal fraction on the Mode I interlaminar resistance and tensile properties of flax fibre/epoxy laminates stitched with flax yarns and cotton threads. Interestingly, both stitch materials caused a similar decrease in tensile properties of the composites, but the presence of flax yarns improved the interlaminar fracture toughness by at least 10%. The results highlighted the need to optimize the areal fraction of stitch to offset the increases in interlaminar fracture toughness and the decrease in tensile properties. The same authors in (Ravandi et al. 2017) investigated the effect of through-the-thickness natural fibre stitches (twistless flax yarn and twisted cotton thread) on the response to low velocity impacts of epoxy laminates reinforced with woven flax fabrics (Fig. 3). In all woven flax fibre composites, cross-shaped cracks were displayed in both the front and the

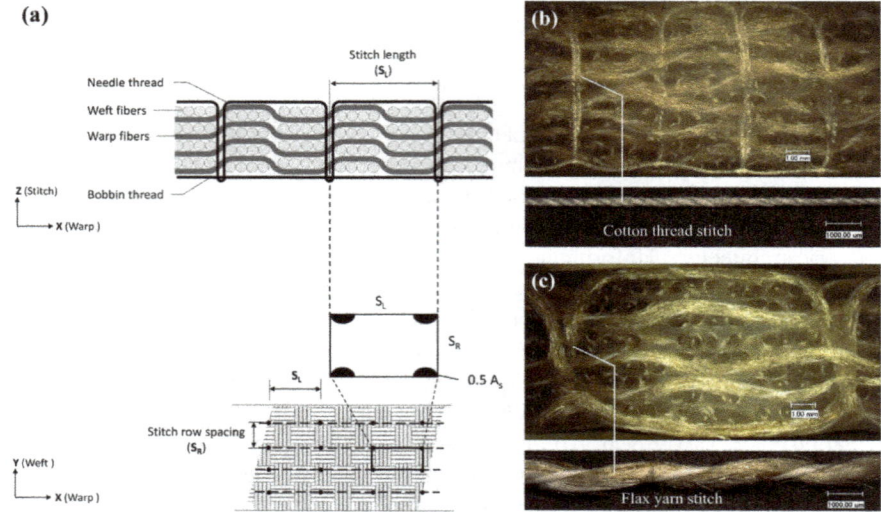

Fig. 3 **a** Schematic view of a stitched preform and definition of stitch parameters; a cross-section of **b** cotton thread, and **c** flax yarn stitched flax fibre composite (reprinted with permission from Ravandi et al. (2017))

rear surfaces of the specimens, and no visible differences were detected between unstitched and stitched specimens, with the exception that crack lengths were longer for the stitched specimens. It is supposed that these cracks were caused by defects originated from the stitches, such as fibre fracture, crimping of in-plane fibres, as well as resin-rich spots. Delamination was only detected in unstitched cross-ply laminates $[0/90]_{4s}$.

Prior to the delamination, in-plane fibre dominated breakages occurred, because of the high interlaminar toughness and relatively poor in-plane strength of the woven flax fibre lamina. For non-perforating impacts, the ratio of the absorbed energy and the kinetic impact energy for the stitched woven composites was about 12–18% higher than that of the unstitched woven laminates, ascribed to the defects generated by stitching. Stitching promoted the development of in-plane cracks because delamination was not the dominant damage mode in woven laminates. The matrix-rich areas located between the stitch loops are characterized by a lower resistance to cracking that requires a lower force during impact to initiate damage. The authors concluded that flax yarn stitching of woven flax laminates did not improve the structural behaviour of the composites in response to low-velocity impacts. In natural fibre composites, fibre failure and matrix cracking have been recognized as the governing mechanisms during a transverse impact, contrary to what happens in standard glass or carbon fibre reinforced composites.

In addition to the fabric architecture, the stacking sequence is another key design parameter for conventional fibre reinforced composites in order to face impact loading, and many investigations are available in this field (Hitchen and Kemp

1995; Hosur et al. 1998; Aktaş et al. 2013; Riccio et al. 2014; Hazzard et al. 2017; Caminero et al. 2017). As a general comment, the damage resistance can be improved by reducing the mismatch in stiffness between neighbouring plies (Caminero et al. 2017), therefore quasi-isotropic laminates are usually characterized by better performance in terms of damage resistance. The use of dispersed configurations with small mismatch angles can result in a superior response in terms of indentation, dissipated energy and residual compressive strength (Sebaey et al. 2013a, b). It is also worth mentioning that the damage modes are influenced by the ply thickness. In thin ply laminates the failure originates in the bottom plies due to bending stresses, which then propagate through the thickness up to the impacted face. Thick laminates are characterized by matrix cracks that emanate from the front side due to the high contact stresses and subsequently travel toward the bottom plies, leading to the typical pine tree damage pattern.

The influence of the stacking sequence on the low-velocity impact and damage tolerance of natural fibre-reinforced composites was investigated by Li et al. (2020). The authors considered three different stacking sequences, i.e. cross-ply $[0/90]_{6s}$, quasi-isotropic $[0/45/90/-45]_{3s}$ and multi-directional ply $[0/30/60/90/-30/-60]_{2s}$. For comparison purposes, a similar glass fibre reinforced composite with the quasi-isotropic configuration was manufactured and tested. The authors highlighted differences with the results usually found for synthetic laminates (Caminero et al. 2017). In particular, the cross-ply composite showed the highest peak load, followed by the quasi-isotropic (by 6%) and multi-directional ply composite (by 7.1%). When compared with the glass fibre composite, synthetic composite displayed the highest peak load indicating the highest impact resistance. This different behaviour resulted in a higher penetration resistance of the glass fibre composite compared to the flax fibre reinforced laminates, among which the best performing laminates were the cross-ply ones. The authors also discussed the damage development and failure mechanisms of the composites. Cross-ply laminates displayed a delamination with the shape of a cross on the impacted surface coupled with two crossed cracks on the rear face, while the other two configurations exhibited a circular indentation on the front surface and an extended delaminated area on the rear surface. Cross-ply laminates absorbed lower energy due to the limited extension of the delaminated area. In cross-ply laminates, once the cracks originate on the back surface, they propagate through the thickness, but they are hampered by the flax fibres oriented at 90° that lie in the neighbouring layer. In this way, a new intra-laminar crack in the fibre direction is triggered and propagate in the adjoining layer. It can be summarized that the interaction between two perpendicular intra-laminar cracks stemming from neighbouring plies is negligible, hence resulting in lower inter-laminar stress and limited delamination. This is not the case in multi-directional or quasi-isotropic laminates, where intra-laminar cracks stemming from adjacent layers display only slightly different angles, and this results in a stronger interaction. This suggests that the through-thickness intra-laminar cracks can readily change their direction and spread through the interface, inducing extensive delamination between the two neighbouring intra-laminar cracks. Synthetic composites with a quasi-isotropic configuration exhibited a different damage pattern, characterized by an overall lower damaged area mainly in the form of a small circular delaminated area. During low velocity impact events, the main drawback of natural fibre composites is related to their inherent lower mechanical properties compared to synthetic fibres, which promote the initiation of transverse cracks at the back surface due to the high bending stresses. These cracks then propagate transversely and through the thickness direction. On the contrary, the glass fibre composites show higher stiffness and strength that result in a lower deflection not able to fracture glass fibres, and delamination is the preferred damage mode. As regards the damage tolerance, the compression after impact behaviour of the cross-ply flax fibre composites outperformed those belonging to the other stacking sequences, namely $[0/30/60/90/-30/-60]_{2s}$ and $[0/45/90/-45]_{3s}$.

In another study, Sy et al. (2018) compared the impact performance of flax fibre reinforced laminates with two different stacking sequences, namely a symmetric unidirectional flax/epoxy laminate $[0]_{8S}$ and a symmetric cross-ply flax/epoxy laminate $[0/90]_{4S}$. In this case, the instrumented drop-weight impact testing machine was replaced with a modified Charpy impact pendulum apparatus. In line with experimental results on synthetic fibre reinforced composites (Ahmad et al. 2015), cross-ply flax/epoxy laminates absorbed higher energy compared to unidirectional flax/epoxy laminates. The latter showed a penetration energy of 10 J, while the former was not completely penetrated even after a 30 J-impact. The two different stacking sequences were characterized by different visible damages on the front and back faces (Figs. 4 and 5). Unidirectional laminates were characterized by a similar damage on the front and rear faces, though it was more pronounced on the rear side. This damage consisted in a critical longitudinal crack starting from the centre of the specimen, a smaller transverse crack running across the longitudinal one decorated with two additional small cracks at its edges (Fig. 4).

On the contrary, impacted and rear faces of cross-ply laminates displayed different damage patterns (Fig. 5). The impacted face damage consisted of a central longitudinal crack combined with a delaminated region with the shape of a butterfly coming from the outermost layer close to the impact location, whereas the back-face damage was made of two cross-shaped matrix cracks. The cross-ply configuration featured the absence of delaminations in the rear face, contrary to what reported in synthetic fibre composites (Namala et al. 2014), due to lower fibre strength compared to synthetic fibres.

In conventional laminates, is the matrix resistance in the rear face that governs the damage development, which is basically in the form of delamination and matrix cracking. In flax/epoxy laminates, the damage produced on the back-face is governed by the fibre properties that are poor, this leading to fibre breakage without significant delaminations. Figures 6 and 7 show an illustrative view of the damage development in through-the-thickness direction for unidirectional and cross-ply flax/epoxy laminates, respectively. In cross-ply laminates the delamination on the impacted side is generated by the interlaminar shear stresses at 0°/90° interfaces, but no delamination is possible in the rear face because the lower strength of flax fibres governs the failure.

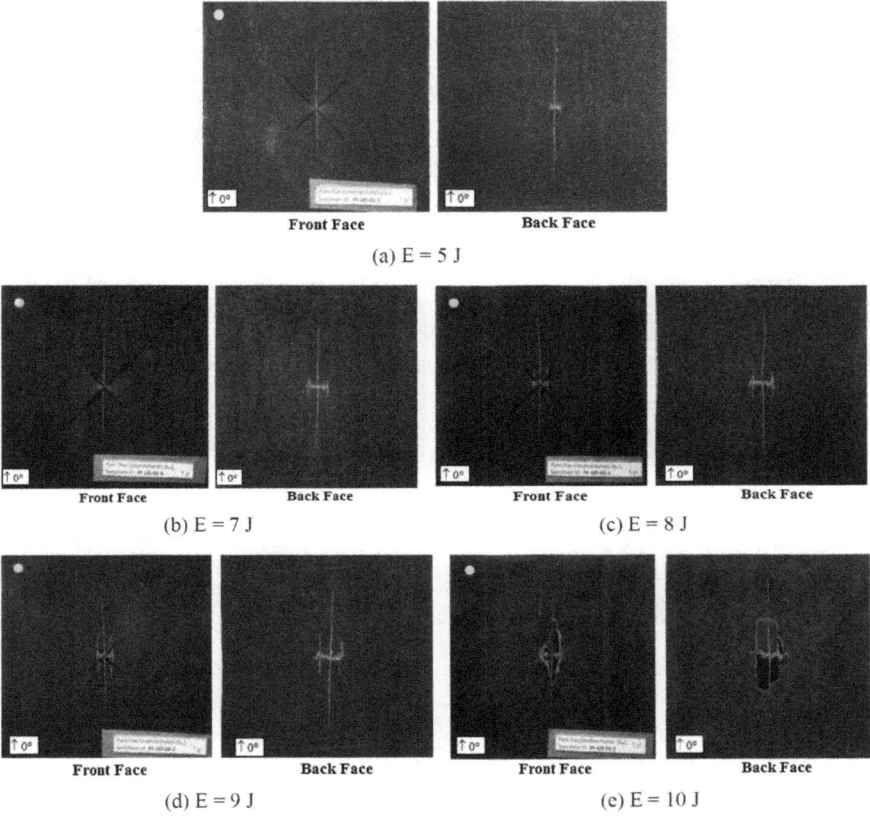

Fig. 4 Front and back face damage on the unidirectional flax/epoxy composite (reprinted with permission from Sy et al. (2018))

Also in quasi-isotropic flax/epoxy composites ($[0/90/45/-45]_{2s}$), Liang et al. (2015) reported fracture mechanisms controlled by the development, in the specimen's rear face, of intra-laminar transverse cracks and a macrocrack due to the failure of the flax fibres (Fig. 8), once again pointing out the prominent role played by the inherent low mechanical properties of natural fibres. After a CAI test, the authors reported a decrease in compression strength of around 30%, similar to that of epoxy composites reinforced with glass fibres (17–34%) (Icten et al. 2013).

Despite the poor mechanical properties of natural fibre composites compared to synthetic counterparts, they still can be used as inexpensive and sustainable energy absorbing structures, even in the ballistic regime (Wambua et al. 2007). Meredith et al. (2012) compared the specific energy absorption (SEA) of jute (plain weave), flax (satin weave) and hemp (chopped strand hemp mat)-based/epoxy composites with that of carbon (plain weave) fibre/epoxy composites. The authors manufactured by vacuum assisted resin transfer moulding cones with an angle of 15°. This geometry

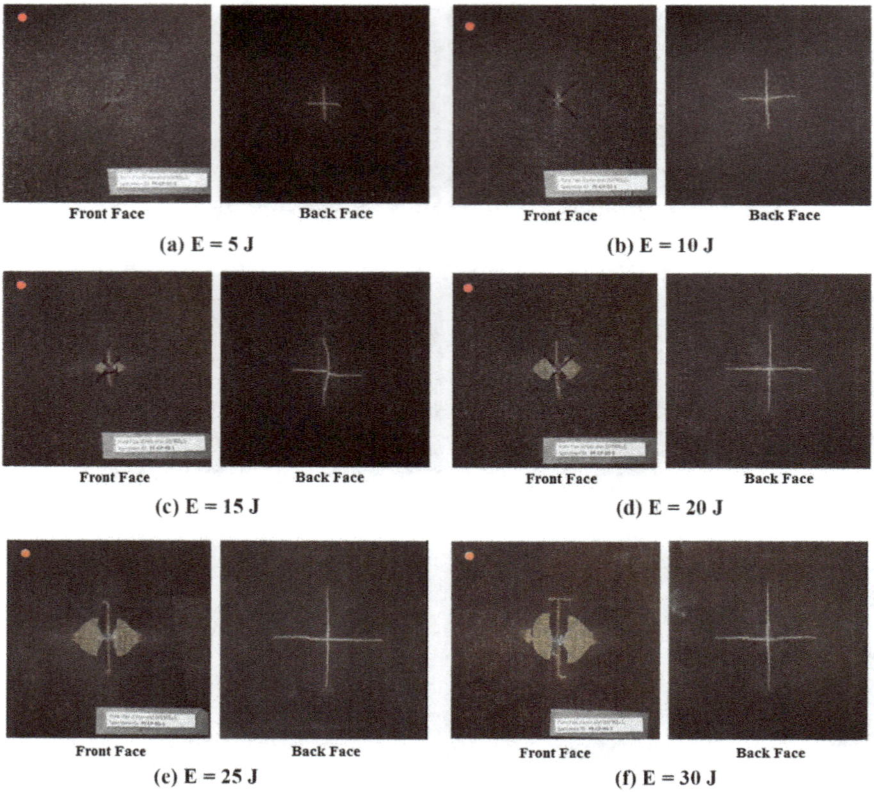

Fig. 5 Front and back face damage on the cross-ply flax/epoxy laminate (reprinted with permission from Sy et al. (2018))

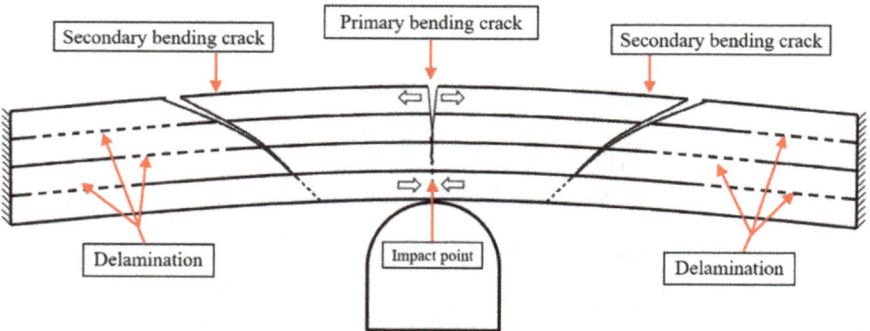

Fig. 6 Schematic of a typical cross section damage on unidirectional flax/epoxy laminate (reprinted with permission from Sy et al. (2018))

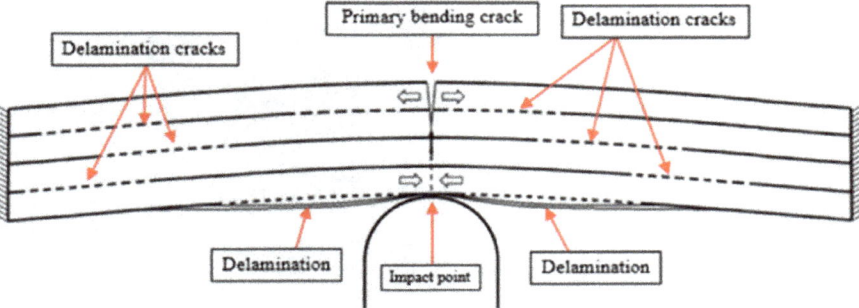

Fig. 7 Schematic of a typical cross section damage on a cross-ply flax/epoxy laminate (reprinted with permission from Sy et al. (2018))

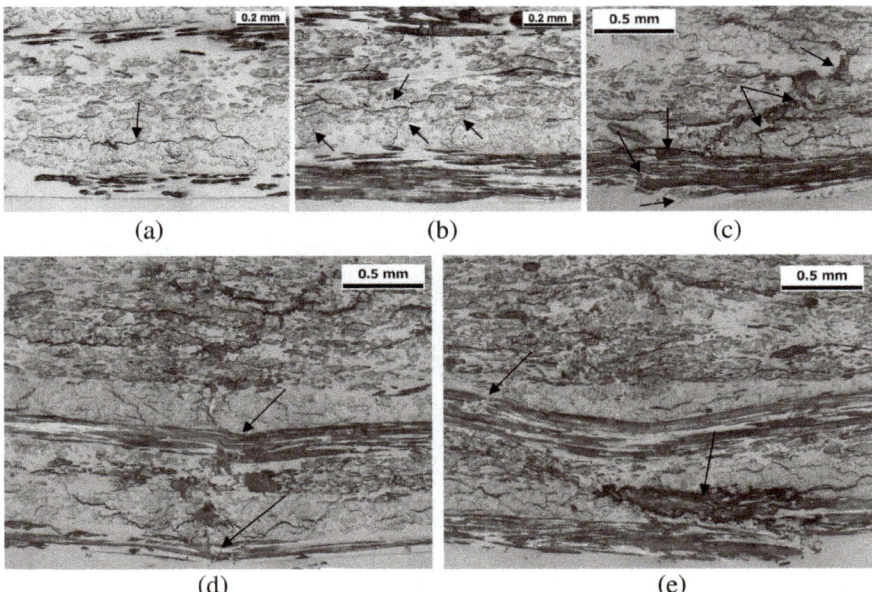

Fig. 8 Damage evolution in quasi-isotropic specimens impacted at different energy levels: **a** 2 J, **b** 4 J, **c** 6 J, **d** 8 J, **e** 10 J. Arrows point to the cracks (reprinted with permission from Liang et al. (2015))

was selected to ensure vertical crush resistance while a wall thickness of 3 mm helped avoiding instability issues during the progressive crushing. Samples were impacted with impact velocities ranging from 3.78 up to 6.70 m/s (Fig. 9). Both woven flax and jute specimens displayed a brittle fracture. In contrast, the non-woven hemp exhibited a progressive collapse with no evidence of terminal longitudinal crack formation upon crush initiation, which resulted in a higher specific energy absorption value (54.3 J/g)

Fig. 9 Comparison of all
samples at different time
points during impact tests
(reprinted with permission
from Meredith et al. (2012))

similar to that of carbon fibre (55.7 J/g) composites, whereas woven flax displayed
a specific energy absorption of 48.5 J/g and woven jute 32.6 J/g.

2.3 Effect of Fibre Hybridization on Impact Behaviour

Hybridization is a common technique used to improve the properties of composite
materials, also when impact performance is taken into consideration (Sevkat et al.
2009; Swolfs et al. 2014, 2018; Bandaru et al. 2016). For natural fibres, hybridization is a significant opportunity to achieve a sufficient mechanical performance
for semi-structural applications whereas reducing the carbon footprint of synthetic
composite materials (Santulli 2007; Jawaid and Abdul Khalil 2011; Dong 2018;
Ravishankar et al. 2019). Glass fibre is the most common synthetic fibre used in
combination with natural fibres, and these hybrid composites are usually characterized by improved mechanical properties, reduced property variability and moisture
sensitivity (Almeida Júnior et al. 2012; Atiqah et al. 2014; Fiore et al. 2016).

An important parameter, when considering impact behaviour of hybrid composites, is the positioning of the layers (Safri et al. 2018). In particular, two main types

of hybridization are possible: (i) the intraply hybridization, when a ply is formed by mixing yarns of two different fibres, (ii) and the interply one, where plies belonging to two different reinforcements are stacked. In the typical configuration, i.e. the interply, the dispersion is completely determined by the lay-up. The positioning of the layers in an interply hybrid composite is of utmost importance, as this affects the flexural stiffness, strength, and the resulting damage mechanisms. Santulli et al. (2005) showed that for E-glass/flax hybrid epoxy composites, the sandwich configuration with glass fibre facesheets and flax fibre core represents the best configuration for improving the impact resistance. Also Ahmed et al. (2007) used a sandwich-like configuration in jute/glass hybrid composites. Jute composites showed higher absorbed energies than jute–glass hybrid laminate, but poorer damage resistance and tolerance that were improved by glass fibre hybridization. Shahzad (2011) investigated the effect of hybridization of hemp fibres with glass fibres on the impact properties of hybrid composites. Two sandwich-like configurations were manufactured: hemp skin and glass core, and glass skin and hemp core. From the results, the replacement of only 11% by volume of hemp fibres with glass fibres resulted in a remarkable increase in residual strength and stiffness of hybrid laminates compared to hemp fibre composites, while the impact damage tolerance of the configuration with glass skins and a hemp core was better than that of composites with hemp facesheets and a glass core, ascribed to the greater mechanical properties of glass fibres. Similar conclusions were achieved by Fragassa et al. (2018), who reported a better impact performance for hybrid composites featuring a flax core sandwiched between basalt fibre facesheets without a non-significant increase in weight.

Glass and basalt fibres (Petrucci et al. 2015; Dhakal et al. 2015; Papa et al. 2018; Ricciardi et al. 2019) have been widely used to increase the impact performance of natural fibre composites, while the combination with carbon fibres has received comparably less attention. This is due to the marked difference in cost and stiffness between natural and carbon fibres (Noorunnisa Khanam et al. 2010; Fiore et al. 2012; Dhakal et al. 2013; Flynn et al. 2016). In this regard, carbon/natural fibre combination has some potential because natural fibres might introduce different modes of damage propagation and energy dissipation, with a view to alleviating the inherent limited toughness of carbon fibre composites. Al-Hajaj et al. (2019) assessed the impact response of an hybrid composite displaying a sandwich structure with woven carbon fibres and flax fibres (i.e., unidirectional and cross-ply) in an epoxy matrix. This particular configuration, as previously mentioned, was chosen because it was supposed that the presence of two carbon/epoxy plies as facesheets was enough to enhance the impact properties compared to neat flax/epoxy laminates. As a general conclusion, both hybrids displayed better resistance to penetration compared to non-hybrid composites, while the composites with the cross-ply flax core performed somewhat better featuring lower absorbed energy, higher penetration energy, smaller crack lengths, smaller indentation depths and smaller damage areas. The penetration energy for the hybrid laminate with the cross-ply flax core was equal to 40 J, higher than neat unidirectional flax fibre (10 J) and neat cross-ply flax fibre (25 J) composites investigated in (Sy et al. 2018). The cross-ply architecture allowed more energy absorption by restraining matrix crack initiation and propagation through the thickness. Fibre bridging at cross-over points between warp and weft yarns might distribute the stress, thus increasing the energy absorption while retaining the structural integrity (Naik et al. 2000).

Hybridization affects the development of damage inside an impacted laminate as well as its damage tolerance and this demands an in-depth analysis of the stacking sequence. Sarasini et al. (2016) addressed this issue by impacting carbon/flax hybrids with two different stacking sequences, i.e., FCF ($[(0_2/90_2)^F/(0_2/90_2)^C/0^C]_S$) and CFC ($[(0_2/90_2)^C/(0_2/90_2)^F/0^F]_S$). The results suggested that the requirements of flexural and impact loading are different in terms of the respective positioning of flax and carbon plies. In particular, the presence of flax fibres as facesheets is useful for impact performance because they restrain crack propagation but not for flexural properties. In terms of residual properties, FCF configuration displayed a better flexural strength and a similar stiffness when compared with laminates made of carbon fibres. The role of outer flax layers in reducing impact damage can be appreciated in Fig. 10, which shows micro-CT scans for the different laminates after a 10 J-impact. While pure carbon (C) fibre reinforced composites displayed the well-known pine tree damage pattern that includes shear cracks, bending cracks and extensive delaminations, pure flax (F) specimens exhibited a damage pattern mostly based on severe transverse matrix cracks, both inter- and intra-yarn in nature. These damage modes were markedly affected by hybridization. In CFC laminates, both carbon facesheets and flax core appear to be damaged in the form of delaminations and bending cracks mainly found in the carbon facesheets. The flax core showed delaminations not only

Fig. 10 Micro-CT scans for the different carbon/flax configurations after a 10 J-impact (reprinted with permission from Sarasini et al. (2016))

at the 0°/90° interfaces but also at the flax/carbon interfaces. In FCF composites, only the carbon core showed the typical pine tree damage pattern, while the flax skins were not extensively damaged apart from bending cracks located in the 0° lower plies. The compliant flax skins restrained the propagation of cracks originated in the carbon core, which resulted in a higher damage tolerance after impact.

Compared to the most common interply hybrid composites, intraply hybrid composites have received limited attention. Despite contrasting results in literature (Pegoretti et al. 2004; Wang et al. 2008), the intraply hybrid composites should offer improved resistance to crack propagation during an impact event. Zhang et al. (2018) investigated how different hybrid configurations made of interlayer and intralayer warp-knitted fabrics with carbon and glass fibres can affect the low-velocity impact performance. The intralayer hybrid showed smaller peak load and higher damage area at the same hybrid ratio and level of impact energy compared to the intralayer config-uration, thus pointing out that a better impact resistance can be obtained by using an intralayer hybridization. This strategy can be exploited for supporting the introduc-tion of natural fibres in at least semi-structural applications (Audibert et al. 2018). Recently Sarasini et al. (2019) proposed a new hybrid intraply woven fabric based on flax and basalt fibres to reinforce epoxy and polypropylene matrices. Laminates with the thermoset matrix compared positively with results available in literature for pure flax laminates. Bensadoun et al. (2017a) addressed the impact behaviour of laminates as a function of flax fibre architectures and matrix types. By taking into account the differences in impact test parameters and similarities in terms of lami-nate thickness (2 mm) and a total fibre volume fraction (0.40), flax/basalt intraply hybrid (15 J) displayed a perforation energy (15 J) about two times higher than that of pure flax laminates, whose values are in the range 5.7–7 J (Bensadoun et al. 2017a). The basalt hybridization was able to counteract the poor transverse strength of flax fibres, hindering the growth of diamond-shaped cracks. Also in this study the superior performance of thermoplastic matrix was confirmed, because the plastic deformation is able to restrain further propagation of the cross-shaped cracks.

3 Conclusions and Future Perspectives

The objective of this chapter was to provide a detailed review on the potential use of natural fibre composites in energy-absorbing applications. The impact response of composites is a topic of significant complexity, as it is governed by multiple factors that are often interrelated. The significant amount of literature available on the subject for synthetic composites do not reflect the same interest in natural fibre composites. In addition, natural fibres display peculiarities in their microstructure and mechan-ical properties compared to synthetic fibres that often deserve specific tests and the results found for synthetic laminates cannot be easily transferred to natural fibre composites. The similarities include the major role played by the matrix toughness on the energy needed to achieve perforation, on the damage resistance and tolerance, which seem largely not affected by the particular fibre architecture in natural fibre composites. For impacts performed at energy levels far from perforation, delamination is not recognized as frequent and detrimental as for synthetic composites. This different behaviour can be ascribed to the high interlaminar fracture toughness of natural fibre composites in combination with some extra energy absorbing mechanisms and the low transverse strength of natural fibres, which tend to localize the damage and trigger the initiation and development of significant cross-like macrocracks. While the effects of fabric architecture and stacking sequence have been adequately addressed, there are other parameters that have received limited attention. In real cases, impact events occur in a random way and can involve different levels of impact energy, impactor masses, geometries, and velocities. Such studies are definitely limited in number for natural fibre composites (Wang et al. 2016; Habibi et al. 2018), and to address this issue extensive experimental campaigns are needed, and possibly the development of methods to simulate them (Sy et al. 2019) in order to limit the number of tests.

Another important area is the assessment of the residual strength of impacted specimens and its correlation with the visible dent depth, which might make easier not only the inspection but also the decision on whether to repair the composite part or to replace it, especially important if the applications of natural fibre composites will abandon the cosmetic field. Damage can be detected by referring to different thresholds that depend on the type of inspection, which are usually classified as follows: BVID (barely visible impact damage), Minor VID (visible impact damage) and Large VID (visible impact damage). These thresholds are connected with the dent depth left by the impactor on the specimen: 0.3–0.5 mm for BVID, 2 mm for Minor VID and 50 mm diameter preformation for Large VID (Talreja and Phan 2019).

A major concern in the field of natural fibre composites is their resistance to hygrothermal ageing, and limited studies are available trying to correlate the effects of temperature and/or moisture on their impact resistance. In this regard the role of fibre hybridization is of paramount importance. Živković et al. (2017) reported how basalt fibre hybridization with flax fibres in a vinylester matrix enhanced the impact behaviour compared to single composites, especially for conditioned samples (35 ppt salt water at 80 °C for 912 h without protection). Fiore et al. (2017) aged under salt fog conditions basalt/jute epoxy composites with two lay-ups (i.e., intercalated and sandwich-like). For the quasi-static properties, the sandwich-like structure enhanced the durability of specimens subjected to salt fog aging conditions, while alternating layers of jute and basalt allowed to reduce the strength loss after accelerated aging. The damage tolerance assessment of aged composites is lacking in the available literature and this gap needs to be bridged.

Another unexploited area but full of potential with a view to increasing the impact resistance of natural fibre composites, deals with the combination of 3D composites with thermoplastic matrices, in order to merge the through-the-thickness reinforcement offered by 3D woven fabric and the toughness of thermoplastic matrices.

References

Ahmad F, Hong JW, Choi HS et al (2015) The effects of stacking sequence on the penetration-resistant behaviors of T800 carbon fiber composite plates under low-velocity impact loading. Carbon Lett 16:107–115. https://doi.org/10.5714/CL.2015.16.2.107

Ahmed KS, Vijayarangan S, Kumar A (2007) Low velocity impact damage characterization of woven jute glass fabric reinforced isothalic polyester hybrid composites. J Reinf Plast Compos 26:959–976. https://doi.org/10.1177/0731684407079414

Aktaş A, Aktaş M, Turan F (2013) The effect of stacking sequence on the impact and post-impact behavior of woven/knit fabric glass/epoxy hybrid composites. Compos Struct 103:119–135

Al-Hajaj Z, Sy BL, Bougherara H, Zdero R (2019) Impact properties of a new hybrid composite material made from woven carbon fibres plus flax fibres in an epoxy matrix. Compos Struct 208:346–356. https://doi.org/10.1016/J.COMPSTRUCT.2018.10.033

Almeida Júnior JHS, Ornaghi Júnior HL, Amico SC, Amado FDR (2012) Study of hybrid intralaminate curaua/glass composites. Mater Des 42:111–117. https://doi.org/10.1016/j.matdes.2012.05.044

Arikan V, Sayman O (2015) Comparative study on repeated impact response of E-glass fiber reinforced polypropylene and epoxy matrix composites. Compos Part B Eng 83:1–6. https://doi.org/10.1016/J.COMPOSITESB.2015.08.051

Atiqah A, Maleque MA, Jawaid M, Iqbal M (2014) Development of kenaf-glass reinforced unsaturated polyester hybrid composite for structural applications. Compos Part B Eng 56:68–73. https://doi.org/10.1016/j.compositesb.2013.08.019

Audibert C, Andreani A-S, Lainé É, Grandidier J-C (2018) Mechanical characterization and damage mechanism of a new flax-Kevlar hybrid/epoxy composite. Compos Struct 195:126–135. https://doi.org/10.1016/J.COMPSTRUCT.2018.04.061

Bandaru AK, Patel S, Sachan Y et al (2016) Low velocity impact response of 3D angle-interlock Kevlar/basalt reinforced polypropylene composites. Mater Des 105:323–332. https://doi.org/10.1016/J.MATDES.2016.05.075

Bensadoun F, Depuydt D, Baets J et al (2017a) Low velocity impact properties of flax composites. Compos Struct 176:933–944. https://doi.org/10.1016/J.COMPSTRUCT.2017.05.005

Bensadoun F, Verpoest I, Van Vuure A (2017b) Interlaminar fracture toughness of flax-epoxy composites. J Reinf Plast Compos 36:121–136. https://doi.org/10.1177/0731684416672925

Bibo GA, Hogg PJ (1996) The role of reinforcement architecture on impact damage mechanisms and post-impact compression behaviour. J Mater Sci 31:1115–1137. https://doi.org/10.1007/BF00353091

Bourmaud A, Beaugrand J, Shaf DU et al (2018) Towards the design of high-performance plant fibre composites. Prog Mater Sci 97:347–408. https://doi.org/10.1016/J.PMATSCI.2018.05.005

Caminero MA, García-Moreno I, Rodríguez GP (2017) Damage resistance of carbon fibre reinforced epoxy laminates subjected to low velocity impact: effects of laminate thickness and ply-stacking sequence. Polym Test 63:530–541. https://doi.org/10.1016/j.polymertesting.2017.09.016

Caprino G, Carrino L, Durante M et al (2015) Low impact behaviour of hemp fibre reinforced epoxy composites. Compos Struct 133:892–901. https://doi.org/10.1016/j.compstruct.2015.08.029

Caprino G, Lopresto V (2001) On the penetration energy for fibre-reinforced plastics under low-velocity impact conditions. Compos Sci Technol 61:65–73. https://doi.org/10.1016/S0266-3538(00)00152-4

Caprino G, Lopresto V, Langella A, Durante M (2011) Irreversibly absorbed energy and damage in GFRP laminates impacted at low velocity. Compos Struct 93:2853–2860

Chen F, Hodgkinson JM (2009) Impact behaviour of composites with different fibre architecture. Proc Inst Mech Eng Part G J Aerosp Eng 223:1009–1017. https://doi.org/10.1243/09544100JAERO451

Chiu CH, Lai MH, Wu CM (2004) Compression failure mechanisms of 3-D angle interlock woven composites subjected to low-energy impact. Polym Polym Compos 12:309–320. https://doi.org/10.1177/096739110401200405

Dhakal HN, Sarasini F, Santulli C et al (2015) Effect of basalt fibre hybridisation on post-impact mechanical behaviour of hemp fibre reinforced composites. Compos Part A Appl Sci Manuf 75:54–67. https://doi.org/10.1016/j.compositesa.2015.04.020

Dhakal HN, Zhang ZY, Guthrie R et al (2013) Development of flax/carbon fibre hybrid composites for enhanced properties. Carbohydr Polym 96:1–8. https://doi.org/10.1016/j.carbpol.2013.03.074

Dittenber DB, GangaRao HVS (2012) Critical review of recent publications on use of natural composites in infrastructure. Compos Part A Appl Sci Manuf 43:1419–1429. https://doi.org/10.1016/j.compositesa.2011.11.019

Dong C (2018) Review of natural fibre-reinforced hybrid composites. J Reinf Plast Compos 37:331–348. https://doi.org/10.1177/0731684417745368

Elias A, Laurin F, Kaminski M, Gornet L (2017) Experimental and numerical investigations of low energy/velocity impact damage generated in 3D woven composite with polymer matrix. Compos Struct 159:228–239. https://doi.org/10.1016/j.compstruct.2016.09.077

Fiore V, Scalici T, Calabrese L et al (2016) Effect of external basalt layers on durability behaviour of flax reinforced composites. Compos Part B Eng 84:258–265. https://doi.org/10.1016/j.compositesb.2015.08.087

Fiore V, Scalici T, Sarasini F, et al (2017) Salt-fog spray aging of jute-basalt reinforced hybrid structures: flexural and low velocity impact response. Compos Part B Eng 116. https://doi.org/10.1016/j.compositesb.2017.01.031

Fiore V, Valenza A, Di Bella G (2012) Mechanical behavior of carbon/flax hybrid composites for structural applications. J Compos Mater 46:2089–2096. https://doi.org/10.1177/0021998311429884

Flynn J, Amiri A, Ulven C (2016) Hybridized carbon and flax fiber composites for tailored performance. Mater Des 102:21–29. https://doi.org/10.1016/j.matdes.2016.03.164

Fragassa C, Pavlovic A, Santulli C (2018) Mechanical and impact characterisation of flax and basalt fibre vinylester composites and their hybrids. Compos Part B Eng 137:247–259. https://doi.org/10.1016/J.COMPOSITESB.2017.01.004

Francesconi L, Aymerich F (2017) Numerical simulation of the effect of stitching on the delamination resistance of laminated composites subjected to low-velocity impact. Compos Struct 159:110–120. https://doi.org/10.1016/j.compstruct.2016.09.050

Gholampour A, Ozbakkaloglu T (2020) A review of natural fiber composites: properties, modification and processing techniques, characterization, applications. J. Mater. Sci. 55:829–892

Grand View Research Inc. (2018) Natural fiber composites (NFC) market size, share and trends analysis report by raw material, by matrix, by technology, by application, and segment forecasts, pp 2018–2024

Greve L, Pickett AK (2006) Delamination testing and modelling for composite crash simulation. Compos Sci Technol 66:816–826. https://doi.org/10.1016/j.compscitech.2004.12.042

Habibi M, Laperrière L, Hassanabadi HM (2018) Influence of low-velocity impact on residual tensile properties of nonwoven flax/epoxy composite. Compos Struct 186:175–182. https://doi.org/10.1016/j.compstruct.2017.12.024

Hazzard MK, Hallett S, Curtis PT et al (2017) Effect of fibre orientation on the low velocity impact response of thin Dyneema® composite laminates. Int J Impact Eng 100:35–45. https://doi.org/10.1016/j.ijimpeng.2016.10.007

Hitchen SA, Kemp RMJ (1995) The effect of stacking sequence on impact damage in a carbon fibre/epoxy composite. Composites 26:207–214. https://doi.org/10.1016/0010-4361(95)91384-H

Hosur MV, Murthy CRL, Ramamurthy TS, Shet A (1998) Estimation of impact-induced damage in CFRR laminates through ultrasonic imaging. NDT E Int 31:359–374. https://doi.org/10.1016/S0963-8695(97)00053-4

Icten BM, Kiral BG, Deniz ME (2013) Impactor diameter effect on low velocity impact response of woven glass epoxy composite plates. Compos Part B Eng 50:325–332. https://doi.org/10.1016/j.compositesb.2013.02.024

Jang BP, Huang CT, Hsieh CY et al (1991) Repeated impact failure of continuous fiber reinforced thermoplastic and thermoset composites. J Compos Mater 25:1171–1203. https://doi.org/10.1177/002199839102500906

Jawaid M, Abdul Khalil HPS (2011) Cellulosic/synthetic fibre reinforced polymer hybrid composites: a review. Carbohydr Polym 86:1–18. https://doi.org/10.1016/j.carbpol.2011.04.043

Koronis G, Silva A, Fontul M (2013) Green composites: a review of adequate materials for automotive applications. Compos Part B Eng 44:120–127. https://doi.org/10.1016/j.compositesb.2012.07.004

Li Y, Zhong J, Fu K (2020) Low-velocity impact and compression-after-impact behaviour of flax fibre-reinforced composites. Acta Mech Solida Sin 1–18. https://doi.org/10.1007/s10338-019-00158-8

Liang S, Guillaumat L, Gning P-B (2015) Impact behaviour of flax/epoxy composite plates. Int J Impact Eng 80:56–64. https://doi.org/10.1016/J.IJIMPENG.2015.01.006

Liu Q, Hughes M (2008) The fracture behaviour and toughness of woven flax fibre reinforced epoxy composites. Compos Part A Appl Sci Manuf 39:1644–1652. https://doi.org/10.1016/j.compositesa.2008.07.008

Meredith J, Ebsworth R, Coles SR et al (2012) Natural fibre composite energy absorption structures. Compos Sci Technol 72:211–217. https://doi.org/10.1016/j.compscitech.2011.11.004

Mouritz AP (2001) Ballistic impact and explosive blast resistance of stitched composites. Compos Part B Eng 32:431–439. https://doi.org/10.1016/S1359-8368(01)00015-4

Mouritz AP (2007) Review of z-pinned composite laminates. Compos Part A Appl Sci Manuf 38:2383–2397. https://doi.org/10.1016/j.compositesa.2007.08.016

Naik NK, Sekher YC, Meduri S (2000) Polymer Matrix Woven Fabric Composites Subjected to Low Velocity Impact: Part I-Damage Initiation Studies. J Reinf Plast Compos 19:912–954. https://doi.org/10.1106/X80M-4UEF-CPPK-M15E

Namala KK, Mahajan P, Bhatnagar N (2014) Digital image correlation of low-velocity impact on a glass/epoxy composite. Int J Comput Meth Eng Sci Mech 203–217

Noorunnisa Khanam P, Abdul Khalil HPS, Jawaid M et al (2010) Sisal/carbon fibre reinforced hybrid composites: tensile, flexural and chemical resistance properties. J Polym Environ 18:727–733. https://doi.org/10.1007/s10924-010-0210-3

Papa I, Ricciardi MR, Antonucci V et al (2018) Impact behaviour of hybrid basalt/flax twill laminates. Compos Part B Eng 153:17–25. https://doi.org/10.1016/J.COMPOSITESB.2018.07.025

Pegoretti A, Fabbri E, Migliaresi C, Pilati F (2004) Intraply and interply hybrid composites based on e-glass and poly(vinyl alcohol) woven fabrics: tensile and impact properties. Polym Int 53:1290–1297

Petrucci R, Santulli C, Puglia D et al (2015) Impact and post-impact damage characterisation of hybrid composite laminates based on basalt fibres in combination with flax, hemp and glass fibres manufactured by vacuum infusion. Compos Part B Eng 69:507–515. https://doi.org/10.1016/j.compositesb.2014.10.031

Pickering KL, Aruan Efendy MG, Le TM (2016) A review of recent developments in natural fibre composites and their mechanical performance. Compos Part A Appl Sci Manuf 83:98–112. https://doi.org/10.1016/j.compositesa.2015.08.038

Ravandi M, Teo WS, Tran LQN et al (2016) The effects of through-the-thickness stitching on the Mode I interlaminar fracture toughness of flax/epoxy composite laminates. Mater Des 109:659–669. https://doi.org/10.1016/j.matdes.2016.07.093

Ravandi M, Teo WS, Tran LQN et al (2017) Low velocity impact performance of stitched flax/epoxy composite laminates. Compos Part B Eng 117:89–100. https://doi.org/10.1016/j.compositesb.2017.02.003

Ravishankar B, Nayak SK, Kader MA (2019) Hybrid composites for automotive applications–a review. J Reinf Plast Compos 38:835–845. https://doi.org/10.1177/0731684419849708

Ricciardi MR, Papa I, Lopresto V et al (2019) Effect of hybridization on the impact properties of flax/basalt epoxy composites: Influence of the stacking sequence. Compos Struct 214:476–485. https://doi.org/10.1016/j.compstruct.2019.01.087

Riccio A, Di Felice G, Saputo S, Scaramuzzino F (2014) Stacking sequence effects on damage onset in composite laminate subjected to low velocity impact. Procedia Eng 222–229

Rong MZ, Zhang MQ, Liu Y et al (2002) Effect of stitching on in-plane and interlaminar properties of sisal/epoxy laminates. J Compos Mater 36:1505–1526. https://doi.org/10.1177/002199830203 6012163

Safri SNA, Sultan MTH, Jawaid M, Jayakrishna K (2018) Impact behaviour of hybrid composites for structural applications: a review. Compos Part B Eng 133:112–121. https://doi.org/10.1016/J.COMPOSITESB.2017.09.008

Sanjay MR, Madhu P, Jawaid M et al (2018) Characterization and properties of natural fiber polymer composites: a comprehensive review. J Clean Prod 172:566–581

Santulli C (2007) Impact properties of glass/plant fibre hybrid laminates. J Mater Sci 42:3699–3707. https://doi.org/10.1007/s10853-006-0662-y

Santulli C, Janssen M, Jeronimidis G (2005) Partial replacement of e-glass fibers with flax fibers in composites and effect on falling weight impact performance. J Mater Sci 40:3581–3585. https://doi.org/10.1007/s10853-005-2882-y

Sarasini F, Tirillò J, D'Altilia S, et al (2016) Damage tolerance of carbon/flax hybrid composites subjected to low velocity impact. Compos Part B Eng 91. https://doi.org/10.1016/j.compositesb.2016.01.050

Sarasini F, Tirillò J, Ferrante L et al (2014) Drop-weight impact behaviour of woven hybrid basalt–carbon/epoxy composites. Compos Part B Eng 59:204–220. https://doi.org/10.1016/j.compositesb.2013.12.006

Sarasini F, Tirillò J, Ferrante L et al (2019) Quasi-static and low-velocity impact behavior of intraply hybrid flax/basalt composites. Fibers 7:26. https://doi.org/10.3390/FIB7030026

Schrauwen B, Peijs T (2002) Influence of matrix ductility and fibre architecture on the repeated impact response of glass-fibre-reinforced laminated composites. Appl Compos Mater 9:331–352. https://doi.org/10.1023/A:1020267013414

Sebaey TA, González EV, Lopes CS et al (2013a) Damage resistance and damage tolerance of dispersed CFRP laminates: Effect of the mismatch angle between plies. Compos Struct 101:255–264. https://doi.org/10.1016/j.compstruct.2013.01.026

Sebaey TA, González EV, Lopes CS et al (2013b) Damage resistance and damage tolerance of dispersed CFRP laminates: effect of ply clustering. Compos Struct 106:96–103. https://doi.org/10.1016/j.compstruct.2013.05.052

Seltzer R, González C, Muñoz R et al (2013) X-ray microtomography analysis of the damage micromechanisms in 3D woven composites under low-velocity impact. Compos Part A Appl Sci Manuf 45:49–60. https://doi.org/10.1016/j.compositesa.2012.09.017

Sevkat E, Liaw B, Delale F, Raju BB (2009) Drop-weight impact of plain-woven hybrid glass–graphite/toughened epoxy composites. Compos Part A Appl Sci Manuf 40:1090–1110. https://doi.org/10.1016/j.compositesa.2009.04.028

Shah SZH, Karuppanan S, Megat-Yusoff PSM, Sajid Z (2019) Impact resistance and damage tolerance of fiber reinforced composites: a review. Compos Struct 217:100–121

Shahzad A (2011) Impact and fatigue properties of hemp-glass fiber hybrid biocomposites. J Reinf Plast Compos 30:1389–1398. https://doi.org/10.1177/0731684411425975

Shyr T-W, Pan Y-H (2003) Impact resistance and damage characteristics of composite laminates. Compos Struct 62:193–203. https://doi.org/10.1016/S0263-8223(03)00114-4

Sutherland LS, Guedes Soares C (2005) Impact on low fibre-volume, glass/polyester rectangular plates. Compos Struct 68:13–22. https://doi.org/10.1016/j.compstruct.2004.02.010

Swolfs Y, Gorbatikh L, Verpoest I (2014) Fibre hybridisation in polymer composites: a review. Compos Part A Appl Sci Manuf 67:181–200. https://doi.org/10.1016/j.compositesa.2014.08.027

Swolfs Y, Verpoest I, Gorbatikh L (2018) Recent advances in fibre-hybrid composites: materials selection, opportunities and applications. Int Mater Rev 1–35. https://doi.org/10.1080/09506608. 2018.1467365

Sy BL, Fawaz Z, Bougherara H (2018) Damage evolution in unidirectional and cross-ply flax/epoxy laminates subjected to low velocity impact loading. Compos Part A Appl Sci Manuf 112:452–467. https://doi.org/10.1016/J.COMPOSITESA.2018.06.032

Sy BL, Fawaz Z, Bougherara H (2019) Numerical simulation correlating the low velocity impact behaviour of flax/epoxy laminates. Compos Part A Appl Sci Manuf 126:105582. https://doi.org/ 10.1016/j.compositesa.2019.105582

Talreja R, Phan N (2019) Assessment of damage tolerance approaches for composite aircraft with focus on barely visible impact damage. Compos Struct 219:1–7

Vieille B, Casado VM, Bouvet C (2013) About the impact behavior of woven-ply carbon fiber-reinforced thermoplastic- and thermosetting-composites: A comparative study. Compos Struct 101:9–21. https://doi.org/10.1016/j.compstruct.2013.01.025

Wambua P, Vangrimde B, Lomov S, Verpoest I (2007) The response of natural fibre composites to ballistic impact by fragment simulating projectiles. Compos Struct 77:232–240. https://doi.org/ 10.1016/j.compstruct.2005.07.006

Wang W, Chouw N, Jayaraman K (2016) Effect of thickness on the impact resistance of flax fibre-reinforced polymer. J Reinf Plast Compos 35:1277–1289. https://doi.org/10.1177/073168441664 8780

Wang X, Hu B, Feng Y et al (2008) Low velocity impact properties of 3D woven basalt/aramid hybrid composites. Compos Sci Technol 68:444–450. https://doi.org/10.1016/j.compscitech. 2007.06.016

Yan L, Chouw N, Jayaraman K (2014) Flax fibre and its composites–a review. Compos Part B Eng 56:296–317. https://doi.org/10.1016/j.compositesb.2013.08.014

Yasaee M, Bigg L, Mohamed G, Hallett SR (2017) Influence of Z-pin embedded length on the interlaminar traction response of multi-directional composite laminates. Mater Des 115:26–36. https://doi.org/10.1016/j.matdes.2016.11.025

Zhang C, Rao Y, Li Z, Li W (2018) Low-velocity impact behavior of interlayer/intralayer hybrid composites based on carbon and glass non-crimp fabric. Materials (Basel) 11. https://doi.org/10. 3390/ma11122472

Živković I, Fragassa C, Pavlović A, Brugo T (2017) Influence of moisture absorption on the impact properties of flax, basalt and hybrid flax/basalt fiber reinforced green composites. Compos Part B Eng 111:148–164. https://doi.org/10.1016/j.compositesb.2016.12.018

A Low Velocity Impact Properties of Hybrid of Pineapple Leaf Fibre and Kenaf Fibre Reinforced Vinyl Ester Composites

A. A. Mazlan, M. T. H. Sultan, N. Saba, A. U. M. Shah, S. N. A. Safri, M. Jawaid, and S. H. Lee

Abstract In this study, Pineapple leaf/kenaf fibre (PALF/KF) reinforced vinyl ester (VE) hybrid composites have been fabricated and tested with low velocity impact (LVI) and compression after impact (CAI) testing. The PALF/KF/VE hybrid composites have been fabricated by hand lay-up technique. For the LVI testing, there were three different level of energy were analysed, 5, 10 and 15 J. The level of energy was determined by the maximum impact force that the specimen can withstand. In the case of CAI testing, the specimen impacted by 5 J of impact energy produced the maximum compression force before it completely breaks down into two parts. The results showed that the hybrid composites impacted with 15 J of impact energy produced the highest peak force during LVI testing but lowest in CAI strength. The hybrid composites impacted with 5 J of impact energy illustrated the lowest peak force but highest in CAI strength. The study of impact properties of PALF/KF/VE hybrid composites was shown in this research work.

Keywords Pineapple leaf fibre (PALF) · Kenaf fibre · Vinyl ester (VE) · Low velocity impact (LVI) · Compression after impact (CAI)

A. A. Mazlan · M. T. H. Sultan (✉) · A. U. M. Shah
Department of Aerospace Engineering, Faculty of Engineering, Universiti Putra Malaysia, 43400 UPM Serdang, Selangor, Malaysia
e-mail: thariq@upm.edu.my

M. T. H. Sultan · N. Saba · A. U. M. Shah · S. N. A. Safri · M. Jawaid
Laboratory of Biocomposite Technology, Institute of Tropical Forestry and Forest Products, Universiti Putra Malaysia, 43400 UPM Serdang, Selangor, Malaysia

M. T. H. Sultan
Aerospace Malaysia Innovation Centre (944741-A), Prime Minister's Department, MIGHT Partnership Hub, Jalan Impact, 63000 Cyberjaya, Selangor Darul Ehsan, Malaysia

S. H. Lee
Laboratory of Biopolymer and Derivatives, Institute of Tropical Forestry and Forest Products (INTROP), Universiti Putra Malaysia, 43400 UPM Serdang, Selangor, Malaysia

© Springer Nature Singapore Pte Ltd. 2021   131
M. T. H. Sultan et al. (eds.), *Impact Studies of Composite Materials*, Composites Science and Technology, https://doi.org/10.1007/978-981-16-1323-4_9

1   Introduction

Nature-based fibre composites have been extensively used in the various application including structural, aerospace, automotive and maritime application due to their biodegrability, environmentally friendly, cheap and lightweight (Ismail et al. 2019). Natural fibres have been the reinforcement from a long time ago (Nadlene et al. 2016). The industries as well as the government are currently moving towards the green technologies due to their advantages over synthetic materials. Nowadays, most of the research work conducted on the natural fibres were based on their specific performances including their mechanical, physical, thermal and chemical performances. There were several parts of the natural fibres that were taken during conducting the research work such as leaves, stalk, bast, stem etc. The implementation of natural fibres in the manufacturing process will eventually reduce the usage of synthetic fibres (glass, carbon, aluminium) and contributed to the preservation of natural fossil resources. Pineapple leaf fibre and kenaf fibre were the example of natural fibres that being widely used as a reinforcement in the composites.

Pineapple, Ananas Comosus is one of the most essential tropical fruit in the world that were abundantly available. Pineapple's leaves being extracted to obtain its fibres to be reinforced with polymer matrix to form a biodegradable composite. The production of pineapple leaf fibre (PALF) has been proved successfully as a reinforcement material in the production of composites (Kalapakdee and Amornsakchai 2014; Nopparut and Amornsakchai 2016; Yuakkul et al. 2015). Kenaf, Hibiscus Cannabius L., one of the annual herbaceous plant most planted in Africa and Asia. In reinforcing fibres for the composites, kenaf fibre (KF) is one of the most natural fibres used as a reinforcement due to its superior mechanical properties (Saba et al. 2015). Besides being biodegradable, KF possess high mechanical properties due to its chemical constituents (56–64% cellulose, 21–35% hemicellulose and 8–14% lignin) (Mazuki et al. 2011). Previous study investigated the impact performance of PALF/KF reinforced with phenolic hybrid composites (Asim et al. 2017, 2018), while another study conducted an impact testing on PALF reinforced VE composite (Mohamed et al. 2014).

In determining the strength of natural fibres composites, it is important to investigate the impact properties of composite materials especially when they are being widely used as aircraft structures because the tendency to encounter impact force is high. Thus, it is crucial to know how these properties work for better application. Flax was hybridized with basalt fibres and tested with drop impact test (Fragassa et al. 2018). Flax reinforced vinyl ester composites undergone drop impact test (Habibi et al. 2019). Hemp fabric reinforced bio-based epoxy composite tested with low velocity impact testing (Scarponi et al. 2016).

In this study, PALF/KF reinforced vinyl ester (VE) hybrid composites have been fabricated and tested with low velocity impact (LVI) and compression after impact (CAI) testing. Previously, There are some research works that conducted on the same materials, PALF/KF but on different matrix which is phenolic resin (Asim et al. 2017, 2018) and high density polyethylene (HDPE) (Aji et al. 2011). The current research work used VE as a matrix. For the LVI testing, there were three different level of energy were analysed, 5, 10 and 15 J. The level of energy was determined by the maximum impact force that the specimen can withstand. In this project, 15 J is the maximum impact force that the hybrid composites can withstand. In the case of CAI testing, the specimen undergone compression testing to determine the maximum compression force that the specimen can withstand before it completely breaks down into two parts. This research work should give some information on how to implement the natural fibres in such application that worked on impact and compressive testing.

2 Materials and Methodology

2.1 Materials

PALF used in the reinforcement is in thin yarn form, with the average thickness of 1 mm. The PALF is chopped into 30 mm length. The chemical components of PALF used in this study are 18.8% hemicellulose, 82.0% cellulose and 12.7% lignin (Siakeng et al. 2018).

Woven KF has average thickness of 0.51 mm. The woven KF used contains 49% of cellulose, 24% hemicellulose and 21% lignin.

The matrix used was a premium standard Bisphenol-A type epoxy VE resin obtained from Sino Polymer Co., LTD.

2.2 Fabrication of PALF/KF/VE Hybrid Composites

In this research work, the fabrication is done by using two layers of woven KF (300 × 300 mm) for each top and bottom layer while random oriented chopped PALF was in the middle layer of the composites. The figure of composite layer is shown in Fig. 1. The hand lay-up method was used to prepare the composites. The ratio of fibre:resin was 30:70 while the respective ratio of resin:hardener was 100:3. The

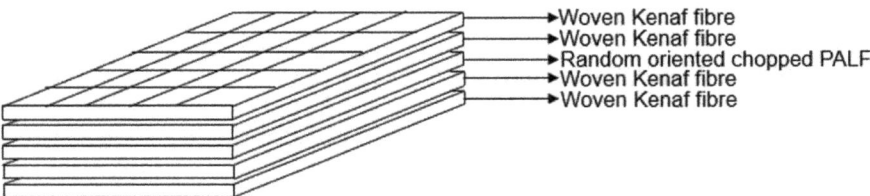

Fig. 1 PALF/KF/VE hybrid composite layer

Table 1 Height of striker

Impact energy (J)	Height of striker (m)
5	0.1
10	0.2
15	0.3

hybrid composites were kept for 24 h before they hardened completely and tested for the next testing.

2.3 Experimental Testing

2.3.1 Low Velocity Impact (LVI) Test

LVI testing was done by using the Imatek IM10 ITS Drop Weight Impact Tester according to ASTM Standard D7136 in aerospace engineering laboratory of UPM. The computer was connected to the instrument that were equipped with data acquisition system in order to record the captured data. A diameter of 10 mm and a mass of 5.101 kg of hemispherical nose striker was used as impactor in this research. The corresponding incident impact velocity was from 1.00 to 2.08 m/s and the incident impact energy varied from 5 to 15 J with an increment of 5 J (adjusted by the height of the impactor). A 150 × 100 mm of specimen dimension were placed on the impact support with a clamped using rubber tip toggle clamps at four points with an identical clamp force. In order to prevent the unintentional multiple impacts on the specimen, a rebound break was used. By using the gravitational energy equation

$$E = mgh$$

where m = mass, g-gravity, h = height in meter (m), and E = energy, the height of the striker was determined based on the value of impact energy as in Table 1.

Three specimens will be tested for each impact energy. This is important in order to obtain constant and average result of the impact energy. All the data being recorded while the table being tabulated. The average values were calculated for each sample.

2.3.2 Compression After Impact (CAI) Test

All the impacted specimens were then further tested with compression testing based on ASTM D7137 in a servo-hydraulic machine (Shimadzu AGX, 300 kN) at the rate of 2.00 mm/min located at the Faculty of Engineering, Universiti Teknologi Malaysia, Kuala Lumpur. Preloading of specimen was applied to ensure all the loading surfaces are in contact as well as to align the plates.

3 Results and Discussion

3.1 Low Velocity Impact (LVI) Test

3.1.1 Force–Time Response

Figure 2 showed the force–time response of PALF/KF/VE hybrid composites subjected to 5, 10 and 15 J of impact energy.

From the curve, the specimen that undergone 15 J of impact energy possessed the highest peak force (2.34 kN) at time of 0.82 ms while the specimen that undergone 5 J of impact energy presented the lowest peak force (1.46 kN) at time of 1.24 ms. The PALF/KF/VE hybrid composites impacted with 15 J of energy exhibits a peak value of maximum force that exceeds the value for the hybrid composites impacted with 5 and 10 J of energy level by at least 63% and 43%, respectively. From 5 to 10 J, the peak force increases 13.7% while from 10 to 15 J, the peak force increases 43.4%. For the hybrid composite impacted with 15 J of impact energy, the stage one lasted at a time of 0.15 ms before the impact force rises again to about 2.27 kN at a time of 0.49 ms. The event of the stage is due to the resistance of bending upper layer of hybrid composite. For the hybrid composite impacted with 10 J of impact energy, the force response after the peak force showed longer plateau than the hybrid composite impacted with 5 J of impact energy. This is due to the PALF that has larger elastic core buckling followed by destruction of its core. The hybrid composite impacted

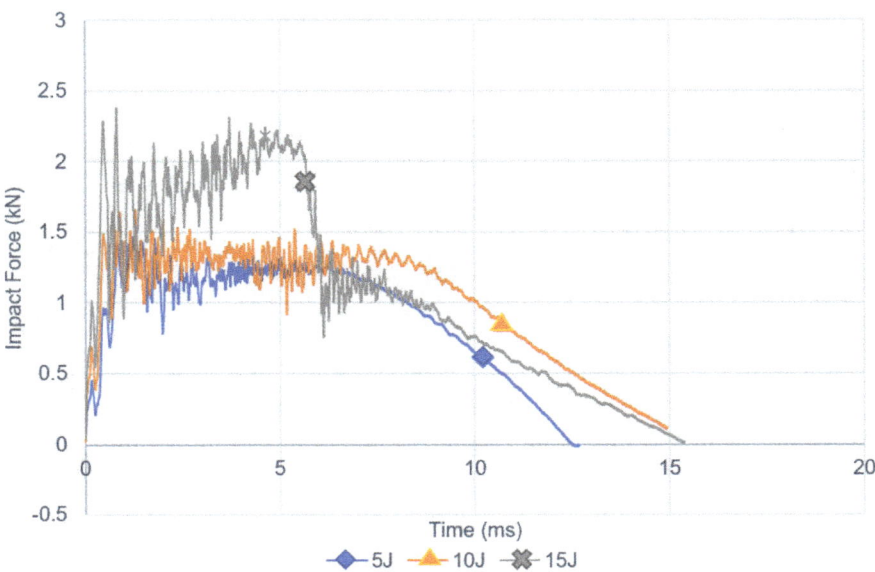

Fig. 2 PALF/KF/VE hybrid composites subjected to 5, 10 and 15 J for force–time response

with 10 and 15 J of impact energy can be compared to each other incorrug force–time response. Figure 3a–c, the impacted specimens showed corrugation angle and core buckling damage of KF, respectively. The repartee fluctuation of force response can be observed because of the external destruction of top layer of hybrid composites. The force curve increases dramatically with increase in impact time until the peak force. Before the force curve reach the peak value, the force–time trace showed a little fall between 400 to 1000 N. This is due to the thin core in thickness which is tendency to buckling. This statement is in lined with (Zangana et al. 2020) who mentioned that the deformation of composite is due to the thin core thickness. With the following increase in impact time, the kinetic energy started to mitigate followed by the gradual decrease in impact force due to the deformation of the hybrid composite and elastic buckling of core. At the time between 0.8 and 9.0 ms, the impact force produced an oscillation due to the disintegration of the KFs' quality at the upper layer of hybrid composites at the impacted area.

Figure 3a–c showed that the elevation of impact energy increases the impact mark for all the hybrid composites. However, the hybrid composite that were impacted with 15 J of impact energy has steeper slopes and cracks than the other two hybrid composites (5 and 10 J). This indicated that the hybrid composites impacted with 15 J of impact energy dispersed the impact force in a larger area compared to other two hybrid composites as it received higher impact energy. The hybrid composite impacted with 5 J of impact energy showed minimal cracks or slope compared the other two. This is due to the height of impactor which is 0.1 m that produced the smallest kinetic energy. Thus, less cracks and slopes produced at the impacted surface area.

Table 2 showed the LVI properties of PALF/KF/VE hybrid composites.

(a)   (b)   (c)

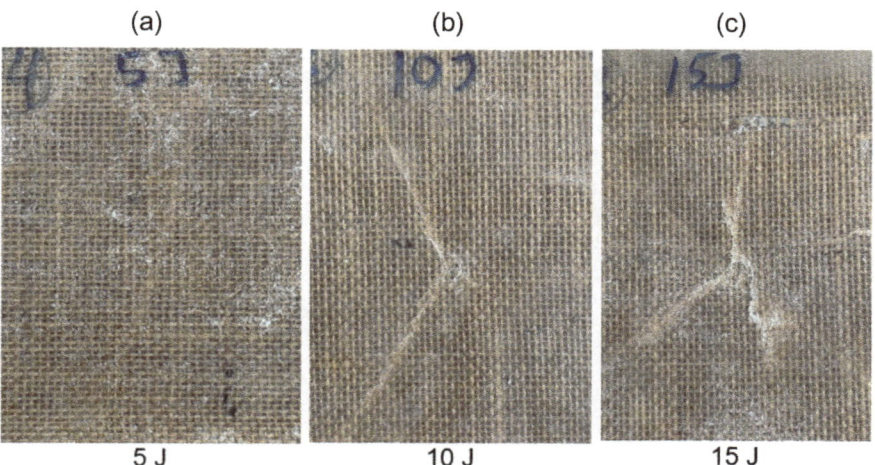

5 J   10 J   15 J

Fig. 3 Captured image of specimen at the end of impact test

Table 2 LVI properties of 5, 10 and 15 J impact energy

Impact energy (J)	Peak force (kN)	Peak deformation (mm)	Absorbed energy (J)	Energy to peak deformation (J)
5	1.46	4.97	3.69	5.3
10	1.66	8.52	8.6	10.53
15	2.38	9.28	14.19	15.65

3.1.2 Force–Displacement Response

The force–displacement graph was presented in the Fig. 4.

The graph of force–displacement showed the closed curve indicates the striker did not manage to penetrate the specimen completely but bounced back after hitting. An open curve happened when the striker completely penetrated the specimen. Thus, from the picture of specimen itself in the Fig. 3a–c, it can be concluded that no specimen experienced penetration, as the incident energy was transferred back to the specimen where the maximum displacement occurred. Maximum displacement recorded when the specimen transfers elastically the excess impact energy to the striker back which leads to bouncing phenomenon between the striker and specimen. From the testing, the specimen can hold up to 15 J before the striker completely penetrated the specimen. The peak deformation of 5, 10 and 15 J impact energy were 4.97 mm, 8.52 mm and 9.28 mm respectively. The curve of specimen that were impacted by 5 and 10 J illustrated quite similar impact force value and showed

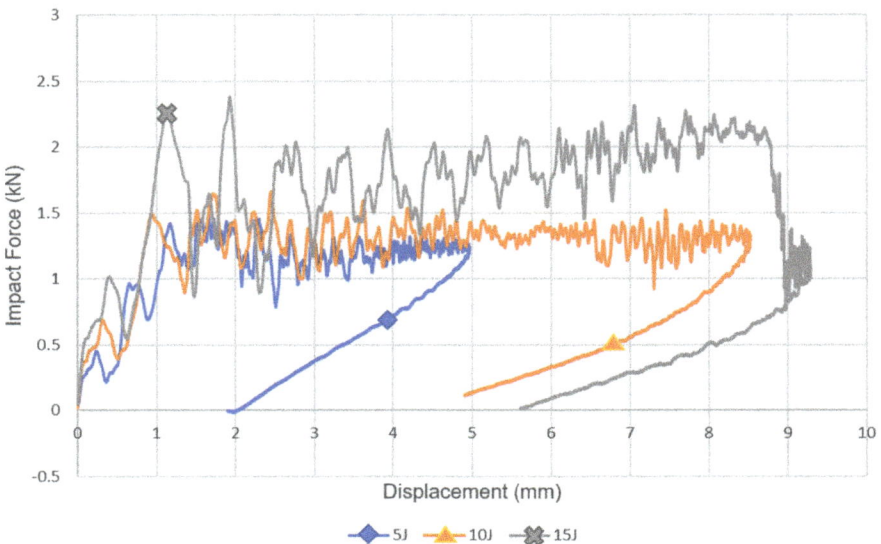

Fig. 4 PALF/KF/VE hybrid composites subjected to 5, 10 and 15 J for force–displacement response

smooth curve rather than the curve of specimen that was impacted by 15 J of impact energy. The curve of specimen impacted by 15 J of impact energy showed larger oscillation and displacement value. The peak deformation recorded associates with the production of cracks produced during impact testing. The sample impacted with 15 J of impact energy showed the highest deformation while the sample impacted with 5 J of impact energy showed the lowest deformation.

3.1.3   Energy–Time Response

The absorbed energy–time responses for all three level of impact energy was presented in Fig. 5.

As the impactor being released on the composites, the kinetic energy in the impactor was being transferred to the laminate composites as soon as the impactor in contact with the laminate composite. The energy began absorbed during the process of delamination, damage and vibration dissipation of the laminate hybrid composites (Tuo et al. 2019). From the curve, the absorbed energy for 5 J impact energy reached 3.69 J while 10 J impact energy showed 8.6 and 14.19 J of absorbed energy recorded by 15 J of impact energy. This is because the height of impactor is the highest among the other samples. The highest height of impactor created the highest kinetic energy, thus create severe interlaminar damage to the composite layers. The surface of KF was severe when the impact energy applied was 15 J. From the Fig. 5, the amount of damage is directly proportional to the amount of energy absorbed.

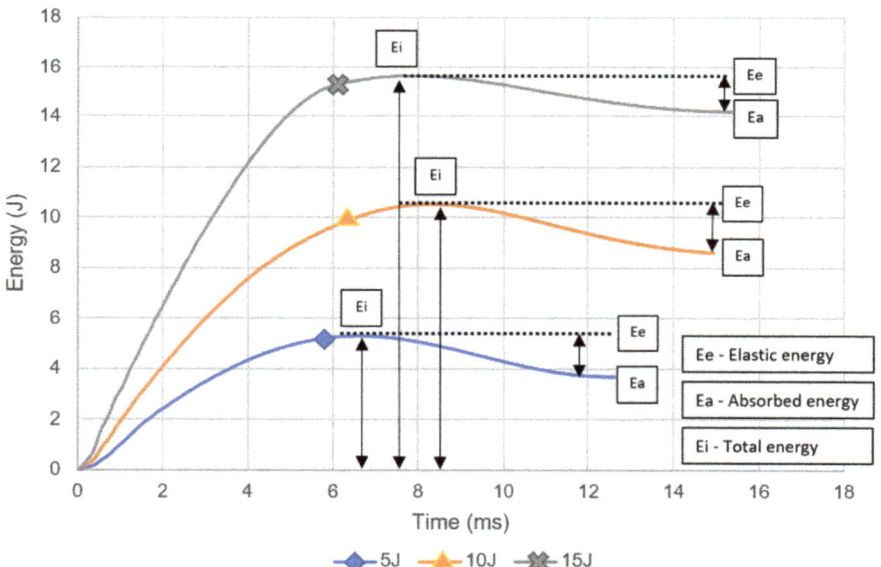

Fig. 5 PALF/KF/VE hybrid composites subjected to 5, 10 and 15 J for energy-time response

3.2 Compression After Impact (CAI) Test

The specimens were further tested with compression after impact (CAI) testing. Figure 6 present the stress versus strain curve of hybrid composites at three different impact energy (5, 10 and 15 J). When the samples are subjected to the compression testing, the compressed samples contributed to the compressive failure, specifically fibre compressive failure and laminated buckling at the back face of the samples. Therefore, due to the impact testing, the fibre and matrix breakage during CAI testing tend to focus on the concentration area. The delamination of the hybrid composites seemed perpendicular to the direction of compression testing. At 5 J of impact energy, the maximum compressive strength recorded was 35.08 kN before the specimen failed, with a displacement of 6.16 mm. At 10 J of impact energy, the maximum compressive strength obtained was 27.08 kN with a displacement of 5.01 mm while 15 J of impact energy produced 18.05 kN prior to failure, with a displacement of 4.19 mm.

Figure 7 showed the ultimate compressive strength of the hybrid composites for different level of energy.

From Fig. 7, the maximum compressive strength recorded was 70.17 MPa for the sample impacted with 5 J, 54.17 MPa for the sample impacted with 10 J and 36.09 MPa for the sample impacted with 15 J. The percentage differences between the compressive strength of sample impacted with 5 J and 10 J is 25.74% while between 5 and 15 J of sample impacted, the percentage differences are 64.14%. From the samples tested, the high impact energy acted on the specimen will produce lower compressive forces and vice versa. This is because due to the LVI testing,

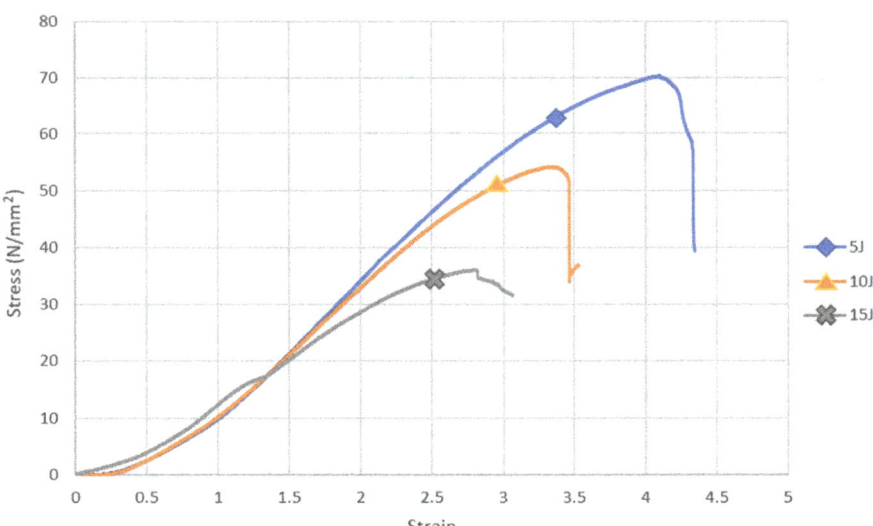

Fig. 6 Stress–strain curve of hybrid composites on CAI testing

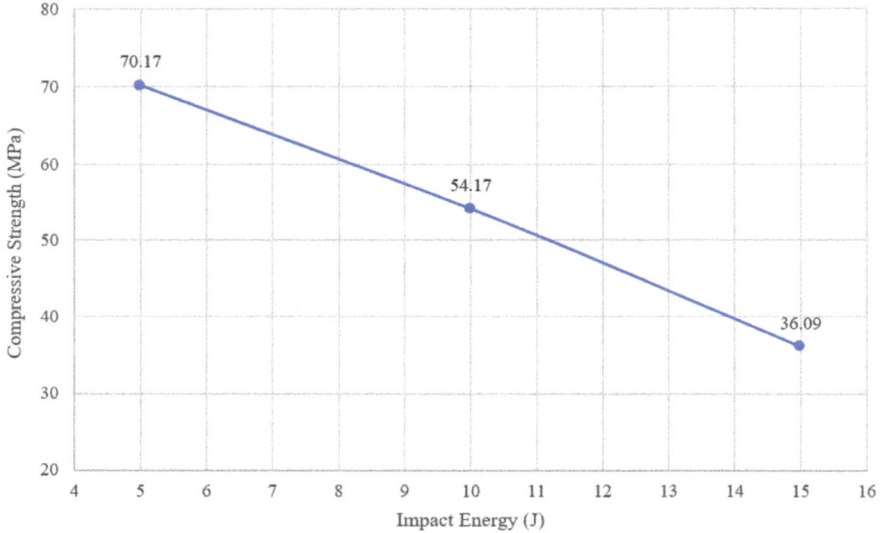

Fig. 7 Compressive strength of PALF/KF/VE hybrid composites

when the energy level increase, the damage of fibre breakage and matrix cracking increase, thus weaken the CAI strength. Another reason is that the high in energy level increase the delamination of composites layer. This result has been found to be in agreement with the previous studies (Rivallant et al. 2014; Selver et al. 2016). The damage on the specimen can be observed by naked eyes as well as the cracks on the samples. It can be understanding that the specimens had previously undergone low velocity impact testing. Thus, the CAI testing had increased the damage area and its cracks. The analysis of CAI properties is important in such application like maritime, aerospace, structural and others.

4 Conclusion

In conclusion, the impact and compressive of PALF/KF/VE hybrid composites were investigated. Based on the discussion above, the conclusion made are as below:

1. The force–time response recorded that the PALF/KF/VE hybrid composites impacted with 15 J energy produced 2.34 kN force.
2. The force–displacement response recorded that when the hybrid composite impacted with 15 J, it produced the largest deformation among the others. The severe delamination in the KF occurred on the PALF/KF/VE hybrid composite impacted with 15 J energy.

3. The energy-time response showed that when the hybrid composites impacted with 15 J of energy, it produced the highest absorbed energy which is 14.19 J due to the highest height, thus produced the highest kinetic energy.
4. For the CAI testing, when the hybrid composite impacted with 15 J, it produced the lowest CAI strength which is 36.09 MPa while the hybrid composite impacted with 5 J of energy produced the highest CAI strength which is 70.17 MPa. It can be noticed that the hybrid composite that was impacted by the highest impact energy will produce the lowest CAI strength.

The current research work that investigated the impact and compressive properties of PALF/KF/VE hybrid composites will provide necessary information needed in real-life application.

Acknowledgements This work is supported by UPM under HICoE grant, 6369107 and Newton Fund, 6300896. The authors would like to express their gratitude and sincere appreciation to the Department of Aerospace Engineering, Faculty of Engineering, Universiti Putra Malaysia and Laboratory of Biocomposite Technology, Institute of Tropical Forestry and Forest Products (INTROP), Universiti Putra Malaysia (HiCOE) for the close collaboration in this research.

References

Aji IS, Zainudin ES, Khalina A, Sapuan SM (2011) Effect of fibre size and fibre loading on tensile properties of hybridized kenaf/PALF reinforced HDPE composite. Key Eng Mater 471–472:680–685. https://doi.org/10.4028/www.scientific.net/KEM.471-472.680

Asim M, Jawaid M, Abdan K, Ishak MR (2017) Effect of pineapple leaf fibre and kenaf fibre treatment on mechanical performance of phenolic hybrid composites. Fibers Polym 18(5):940–947. https://doi.org/10.1007/s12221-017-1236-0

Asim M, Jawaid M, Abdan K, Ishak MR (2018) The effect of Silane treated fibre loading on mechanical properties of Pineapple leaf/Kenaf fibre filler phenolic composites. J Polym Environ 26(4):1520–1527. https://doi.org/10.1007/s10924-017-1060-z

Fragassa C, Pavlovic A, Santulli C (2018) Mechanical and impact characterisation of flax and basalt fibre vinylester composites and their hybrids. Compos B Eng 137:247–259. https://doi.org/10.1016/j.compositesb.2017.01.004

Habibi M, Selmi S, Laperrière L, Mahi H, Kelouwani S (2019) Experimental investigation on the response of unidirectional flax fiber composites to low-velocity impact with after-impact tensile and compressive strength measurement. Compos B Eng 171(October):246–253. https://doi.org/10.1016/j.compositesb.2019.05.011

Ismail MF, Sultan MTH, Hamdan A, Shah AUM, Jawaid M (2019) Low velocity impact behaviour and post-impact characteristics of kenaf/glass hybrid composites with various weight ratios. J Mater Res Technol 8(3):2662–2673. https://doi.org/10.1016/j.jmrt.2019.04.005

Kalapakdee A, Amornsakchai T (2014) Mechanical properties of preferentially aligned short pineapple leaf fiber reinforced thermoplastic elastomer: effects of fiber content and matrix orientation. Polym Testing 37:36–44. https://doi.org/10.1016/j.polymertesting.2014.04.008

Mazuki AAM, Akil HM, Safiee S, Ishak ZAM, Bakar AA (2011) Degradation of dynamic mechanical properties of pultruded kenaf fiber reinforced composites after immersion in various solutions. Compos B Eng 42(1):71–76. https://doi.org/10.1016/j.compositesb.2010.08.004

Mohamed AR, Sapuan SM, Khalina A (2014) Mechanical and thermal properties of josapine pineapple leaf fiber (PALF) and PALF-reinforced vinyl ester composites. Fibers Polym 15(5):1035–1041. https://doi.org/10.1007/s12221-014-1035-9

Nadlene R, Sapuan SM, Jawaid M, Ishak MR, Yusriah L (2016) A review on roselle fiber and its composites. J Nat Fibers 13(1):10–41. https://doi.org/10.1080/15440478.2014.984052

Nopparut A, Amornsakchai T (2016) Influence of pineapple leaf fiber and it's surface treatment on molecular orientation in, and mechanical properties of, injection molded nylon composites. Polym Testing 52:141–149. https://doi.org/10.1016/j.polymertesting.2016.04.012

Rivallant S, Bouvet C, Abi Abdallah E, Broll B, Barrau JJ (2014) Experimental analysis of CFRP laminates subjected to compression after impact: the role of impact-induced cracks in failure. Compos Struct 111(1):147–157. https://doi.org/10.1016/j.compstruct.2013.12.012

Saba N, Jawaid M, Hakeem KR, Paridah MT, Khalina A, Alothman OY (2015) Potential of bioenergy production from industrial kenaf (Hibiscus cannabinus L.) based on Malaysian perspective. Renew Sustain Energy Rev 42:446–459. https://doi.org/10.1016/j.rser.2014.10.029

Scarponi C, Sarasini F, Tirillò J, Lampani L, Valente T, Gaudenzi P (2016) Low-velocity impact behaviour of hemp fibre reinforced bio-based epoxy laminates. Compos B Eng 91(2016):162–168. https://doi.org/10.1016/j.compositesb.2016.01.048

Selver E, Potluri P, Hogg P, Soutis C (2016) Impact damage tolerance of thermoset composites reinforced with hybrid commingled yarns. Compos B Eng 91:522–538. https://doi.org/10.1016/j.compositesb.2015.12.035

Siakeng R, Jawaid M, Ariffin H, Sapuan SM (2018) Thermal properties of coir and pineapple leaf fibre reinforced polylactic acid hybrid composites. IOP Conf Ser: Mater Sci Eng 368(1). https://doi.org/10.1088/1757-899X/368/1/012019

Tuo H, Lu Z, Ma X, Xing J, Zhang C (2019) Damage and failure mechanism of thin composite laminates under low-velocity impact and compression-after-impact loading conditions. Compos B Eng 163:642–654. https://doi.org/10.1016/j.compositesb.2019.01.006

Yuakkul D, Amornsakchai T, Saikrasun S (2015) Effect of maleated compatibilizer on anisotropic mechanical properties, thermo-oxidative stability and morphology of styrenic based thermoplastic elastomer reinforced with alkali-treated pineapple leaf fiber. Int J Plast Technol 19(2):388–411. https://doi.org/10.1007/s12588-016-9132-9

Zangana S, Epaarachchi J, Ferdous W, Leng J (2020) A novel hybridised composite sandwich core with glass, kevlar and zylon fibres – investigation under low-velocity impact. Int J Impact Eng 137:103430. https://doi.org/10.1016/j.ijimpeng.2019.103430

Validation of Experimental Hybrid Natural/Synthetic Composite Laminate Specimen Using Finite Element Analysis for UAV Wing Application

Nur Marini Zainal Abidin, M. T. H. Sultan, Ernnie Illyani Basri, Adi Azriff Basri, and A. U. M. Shah

Abstract In this day and age, there are high demand for the utilisation of hybrid composite materials in industrial and aerospace applications due to its advantages over the common composite material and other materials. In this case, engagement of natural fiber with synthetic fiber to form a hybrid lamination composite to become a partially eco-friendly material had been studied and analysed to pursuit a properties requirement for UAV wing profile. In experimental study, glass and kenaf were fabricated in two variations of GKG and KGK and were tested under tensile test properties. The result obtained was compared by validation of finite element analysis using ANSYS 18.2 workbench. From the validation results, it showed a good agreement in stress and strain but percentage error in deformation due to several factors will be discussed. In term of composite strength, GKG have the maximum stress of 120.89 MPa compared to KGK with 79.787 MPa maximum stress.

Keywords Natural/synthetic laminate composite · Laminate composite wing · Finite element analysis

N. M. Z. Abidin · M. T. H. Sultan (✉) · E. I. Basri · A. A. Basri · A. U. M. Shah
Department of Aerospace Engineering, Faculty of Engineering, Universiti Putra Malaysia, 43400 UPM Serdang, Selangor, Malaysia
e-mail: thariq@upm.edu.my

M. T. H. Sultan
Aerospace Malaysia Innovation Centre (944751-A), Prime Minister's Department, MIGHT Partnership Hub, Jalan Impact, 63000 Cyberjaya, Selangor Darul Ehsan, Malaysia

M. T. H. Sultan · A. U. M. Shah
Laboratory of Biocomposite Technology, Institute of Tropical Forestry and Forest Products, Universiti Putra Malaysia, 43400 UPM Serdang, Selangor, Malaysia

1 Introduction

1.1 Hybrid Composite Laminates

Hybrid composite laminates consisting two or more of types of fibers which together produce desirable properties of strength and modulus fibers in a matrix material. This composite is more advanced than conventional fiber reinforced composite in term of flexibility. Nowadays, engagement of natural fiber in hybrid composite giving a lot of advantages and had been done in several research works. Ramesh and Nijanthan (2016) mentioned that mixing of natural fiber using polymer resins with the synthetic fiber will reduce the cost of production and the harmful destruction. As we known, natural fiber is a renewable source and it can be an alternative solution for environmentally friendly. Salleh et al. (2018) also claimed that natural fiber is a biodegradable and relatively inexpensive compared to synthetic fiber. However, product of natural fiber for structural applications are still limited due to their poor mechanical properties. Compare to synthetic fiber which giving its own advantages such as low weight, high strength, less heat and electrical conductivity and also resistance of chemical agents (Saravanan and Vetrivel 2016). Solving this problem, natural and synthetic fibers are mix together to make the composite hybrid and produce new properties.

In structural applications, fiber-reinforced composite material is form in a lamination of fiber or stacking by a collection of lamina. Hybrid laminate composite term was described as a mixing than one type of fiber in composite laminate. Reddy (2003) claimed that stacking sequence of each fibers and also the orientation can be chosen to achieve desired strength and stiffness. The fiber use in lamination can be continuous or discontinuous, unidirectional, bidirectional, woven or randomly distributed. In this case study, a hybrid composite laminates of E-Glass/Kenaf/E-Glass (G-K-G) fiber and Kenaf/E-Glass/Kenaf (K-G-K) fiber had been fabricated with epoxy resin matrix. The stackup fiber orientation choose is 90/0/90 (Fig. 1). This orientation was applied to both hybrid composite fabricated, G-K-G and K-G-K.

Composite materials which are involved two or more layered materials and have different properties when combining can make this material high strength and superficial with greater manufacturing of complex parts especially in aircraft applications. Some of the product based composite laminate which have complex shapes, will be challenging to analyse or predict the performance of the finished product under real-world working conditions. One of the important component in aircraft using composite laminate as its material is a wing part. Basri et al. (2019b) described that structural analysis in finite element analysis is an effective numerical solution and optimization method in aerospace engineering. In ANSYS workbench, composite lamination material must go through on ANSYS Prepost (ACP) domain where this interface has dedicated tool for composite layup modelling and failure analysis.

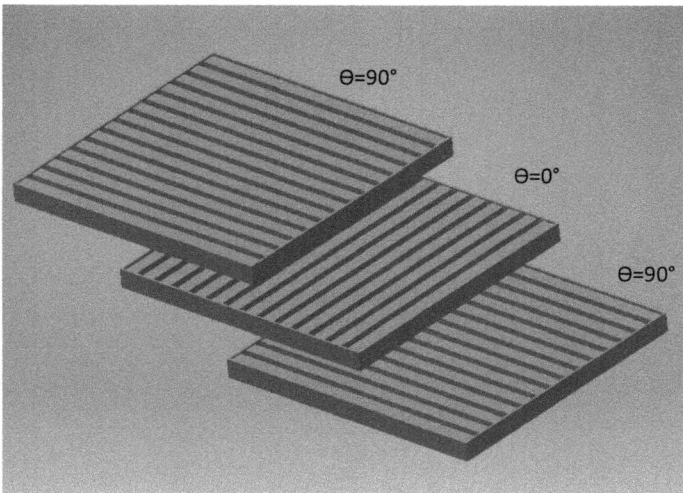

Fig. 1 Fiber orientation of laminate composite

1.2 Finite Element Analysis

A Finite Element Analysis (FEA) is to obtain approximate solutions of Boundary Value Problem using computer technique which based on numerical method solution. A boundary value problem is a solution sought in the domain (or region) of a body subject to the satisfaction of prescribed boundary (edge) conditions on the dependent variables or derivatives (Rao 2004). Three major categories in Boundary Value Problem is Equilibrium problems, Eigenvalue problems and Propagation problems. For example, in aircraft, Equilibrium problem is referring to a static analysis of aircraft wings, fuselage, fins, rockets, etc. while Eigenvalue problems analyse on natural frequencies, flutter and stability on aircraft. Propagation problems is response of aircraft structures to random loads, dynamic response of aircraft (Rao 2004). FEA actually was first developed in the aerospace and nuclear industries where the safety of structures was a main issue to be focused. Nowadays, FEA is widely used in other industries even the simplest products rely on FEA for design evaluation. In the present case study, hybrid composite laminate fabricated was analysed using FEA in order to validated result obtained from the experiment. To perform the simulation, a knowledge of analysis theory of hybrid composite laminate as shown in Fig. 2 is required. The orthotropic elasticity equation (such as Young's Modulus, Poisson's ratio, Shear modulus, Tensile Stress and Strain, Compressive Stress and Strain and Shear Stress and Strain), structural theories (geometry, modelling and ANSYS Composite PrepPost setup), analytical and computational methods to determine the solutions (eg. deformation, stress and strain) and damage or failure theories to predict failure modes and failure loads is the compulsory component or data in finite element analysis (Cook 1995).

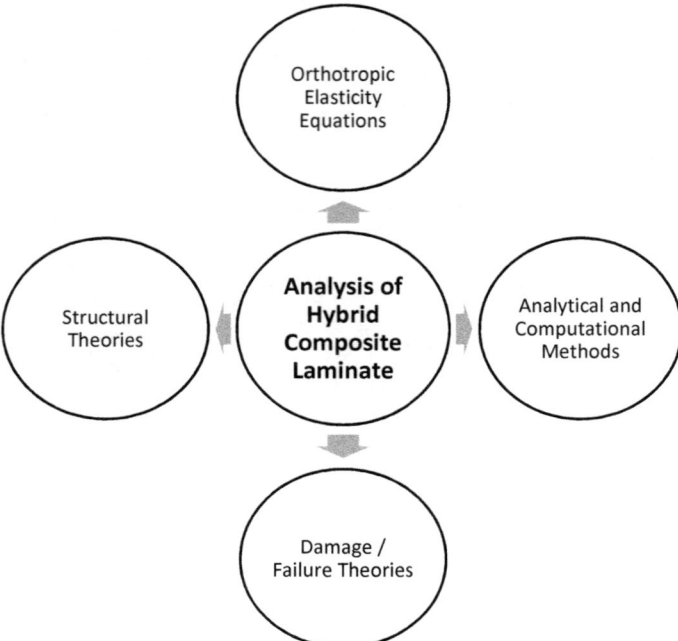

Fig. 2 Analysis of hybrid composite laminate

2 Experimental of Hybrid Natural-Synthetic Laminate Composite

2.1 Material and Methodology

The most widely used and easiest method for laminate composite fabrication is a hand lay-up. For this experiment, the sample size 300 mm width and 300 mm length were prepared. There are two variations of sample were developed which are lamination of Glass-Kenaf-Glass (GKG) and lamination of Kenaf-Glass-Kenaf (KGK). E-Glass and Kenaf fiber were cut properly then measured and recorded all the fiber's weight. In preparation of the mould, a wax was rubbed in a surface of the mould to prevent sticking and enable easy removal of the finished part (Fig. 3).

Next process is mixing epoxy with the hardener (curing agent) with a suitable proportion ratio. Both of the sample were decided using 20:80 proportion ratio of fiber and matrix. The rule of mixture formula calculating as following (Alger 1997):

$$E_c = f E_f + (1 - f)E_m \tag{1}$$

where;

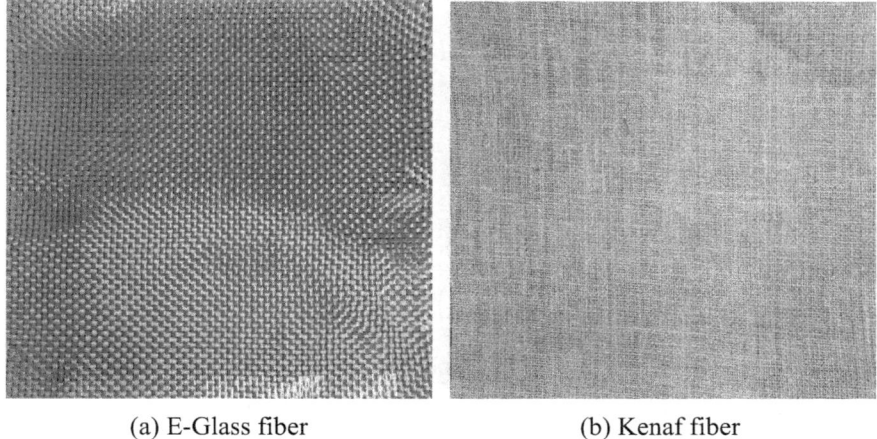

(a) E-Glass fiber   (b) Kenaf fiber

Fig. 3 Material used in fabrication of hybrid natural-synthetic laminate composite

$f = \frac{V_f}{V_f + V_m}$ (the volume of the fibers).

E_f is the material property of the fibers.

E_m is the material property of the matrix.

The mixture of matrix is poured into the mould and spread it uniformly before placed a fiber onto it. Then put the fiber on the matrix and scrub the fiber with the help of brush or roller. The purpose using brush or roller is to remove the extra resin and ensure uniform distribution of resin to whole surface (Biswas and Anurag 2019). The process is repeated for all layers of reinforcements until the required number of layer was achieved. A covering plate of fabrication mould will applied on the top of surface in order to avoid from flying dust falling into the composite laminates sample surface. The sample was left it with standard atmospheric temperature about 24 h before opened and taken out. Then the sample was going to a curing process in the oven with 180° in 2 h (Fig. 4).

Once finished the fabrication process, the samples were cut to a several number of specimens with a specific dimension from ASTM D3039 standard test. The dimension for ASTM D3039 as per Fig. 5.

2.2 Testing and Result

The most important of mechanical testing is determination of material properties of the material which subjected to elasticity, stress and strain data profile. Therefore, the fabricated specimens were undergoing a Tensile Properties Test by followed ASTM D3039 Standard which also known as tension test. ASTM standards were used in

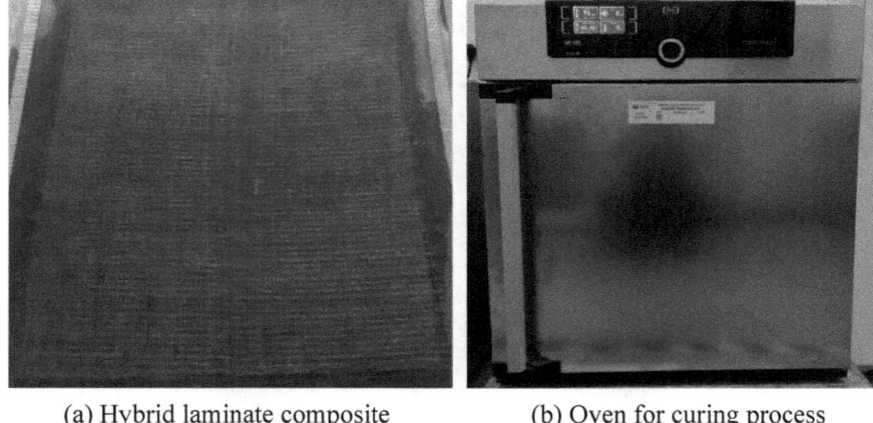

(a) Hybrid laminate composite   (b) Oven for curing process

Fig. 4 Hand layup composite and manufacturing process for hybrid laminate

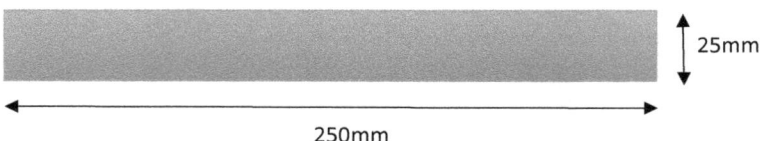

25mm

250mm

Fig. 5 ASTM D3039 standard dimension

producing data of material specifications, research and development, quality assurance and structural design and analysis. When the material was loaded in tension, the maximum strength will be determined from the maximum force before its failure. Result derived from this test including of tensile stress, tensile strain, Modulus Young's, displacement and maximum load which can be initially chosen from the list of mechanical properties in the machine. As usually, before testing the thickness and width each specimen was taken and recorded. Figure 6 showing a 30kN Universal Testing Machine has been used for a tensile test.

During tensile loading, the stress–strain curve was directly generated from the universal testing machine until the composite breaks. As illustrated in Fig. 7a and b, there have a total of 9 samples for GKG and 11 samples for KGK due to defect of GKG sample when in fabrication process. However, both of the curve for each sample which show a brittle material characteristic, demonstrate a minor nonlinear curve until the maximum strength before failures occurred. All the composite samples within its variation also show a good match and similar behaviours. Compared to both of material, a hybrid composite of GKG have a tensile strength and tensile strain more than KGK. The slope of the graph represents the Young's modulus of the composite. In Table 1 is a tabulated data of each composite specimen translated from the stress–strain curve. The average of maximum strength value of GKG is

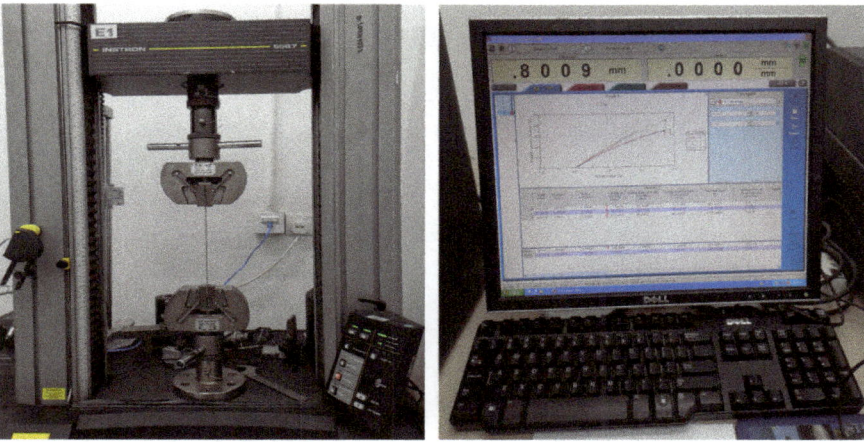

Fig. 6 A 30kN universal testing machine (Instron) and equipment used for the testing

106.648 MPa with an average maximum load of 11,820.874 MPa while KGK have an average maximum strength of 70.17 MPa in average tension force of 7627 MPa.

3 A Development of Laminate Composite Using Finite Element Analysis

Analysis of the composite elastic behaviour in term of material deformation and applied loading conditions were carried out with the help of ANSYS 18.2 Workbench. It is important to note that composite materials analysis quite complex and have a several factors might be affect the analysis (Al-Qrimli et al. 2015). Therefore, validation is required to use a similar ASTM D3039 standard to ensure the analysis and physical testing are accurate. The results obtained must be within 15% differences to consider the parameters is validated (Prasad and Ramachandran 2017). All the variables from experiment were inserted into the data needed from the simulation. For the hybrid composite case, it must be analyse in ANSYS Composite Pre-Post (ACP) 18.2. ACP is an add-on module of ANSYS dedicated to the modelling of layered composite structures. A schematic view of static structural was shown in Fig. 8 below is a subsequent step of the modelling.

3.1 Geometry

The tensile test sample was modelled in geometry step by following ASTM D3039 standard size requirement. The final geometry of the sample shown in Fig. 9.

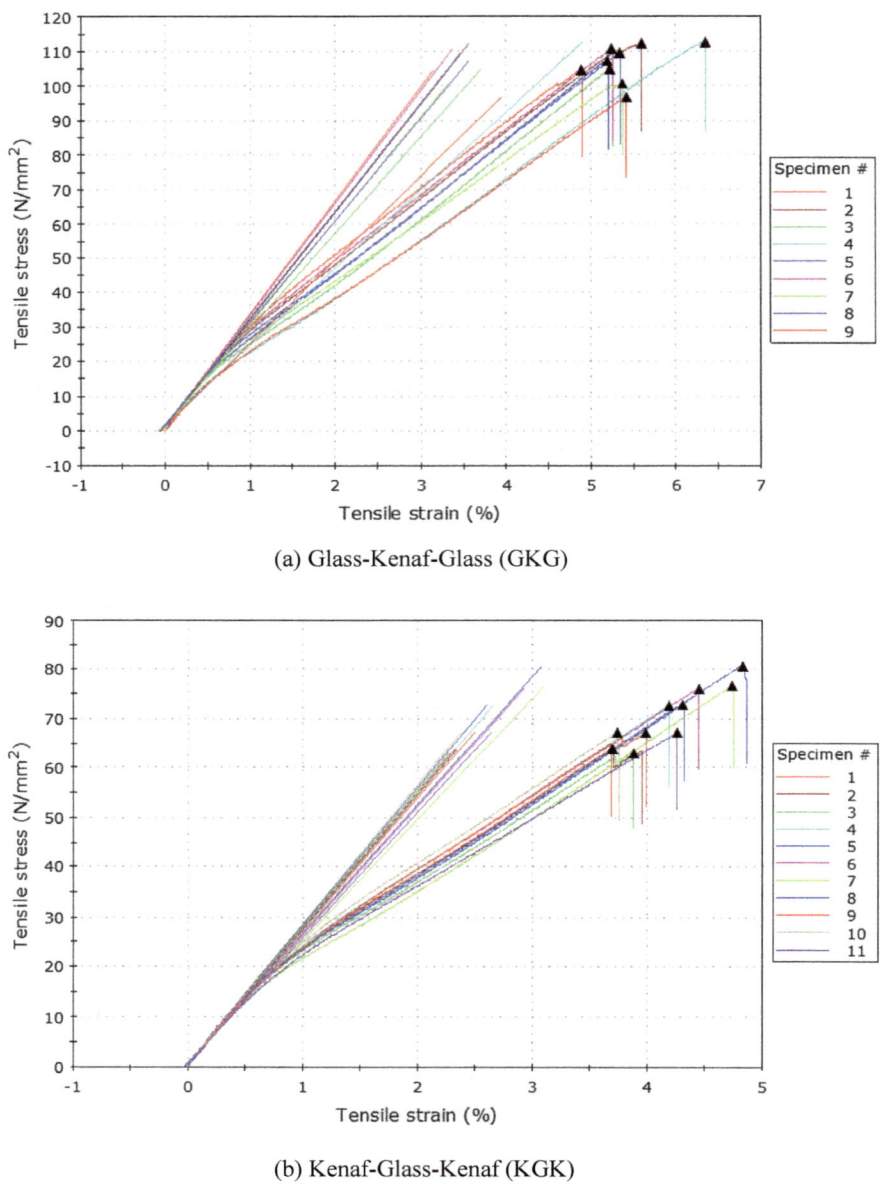

(a) Glass-Kenaf-Glass (GKG)

(b) Kenaf-Glass-Kenaf (KGK)

Fig. 7 Tensile test result for all samples of GKG and KGK laminate composite

Table 1 Summarise of G-K-G and K-G-K tensile test result

No.	Thickness (mm)	Maximum load (N)	T. stress at max load (MPa)	T. strain at max load (mm/mm)	Modulus Young's (MPa)
(a) Tensile test result for Hybrid Glass-Kenaf-Glass lamination composite					
1	4.033	12,397.981	104.623	0.049	3287.366
2	3.843	12,797.241	112.299	0.056	3111.967
3	3.97	12,022.857	104.971	0.052	2769.767
4	3.763	11,941.661	112.705	0.064	2250.534
5	3.85	11,986.564	109.407	0.053	3110.088
6	3.777	12,022.252	110.637	0.052	3244.908
7	4.043	11,416.833	100.805	0.054	2996.642
8	3.973	11,644.989	107.443	0.052	2956.592
9	3.74	10,157.483	96.938	0.054	2415.361
Mean	3.888	11,820.874	106.648	0.054	2904.803
(b) Tensile test result for Hybrid Kenaf-Glass-Kenaf lamination composite					
1	3.733	6976	63.85	0.037	2724.80
2	3.733	6726	63.88	0.037	2700.67
3	3.84	7301	63.11	0.039	2747.00
4	3.84	8022	72.74	0.042	2720.88
5	3.847	7938	72.94	0.043	2775.41
6	3.68	8241	76.1	0.045	2577.91
7	3.72	8029	76.72	0.047	2439.16
8	3.813	9079	80.78	0.048	2597.10
9	3.743	7340	67.19	0.040	2672.15
10	3.733	7429	67.26	0.037	2800.94
11	3.74	6813	67.26	0.043	2525.82
Mean	3.7656	7627	70.17	0.042	2661.98

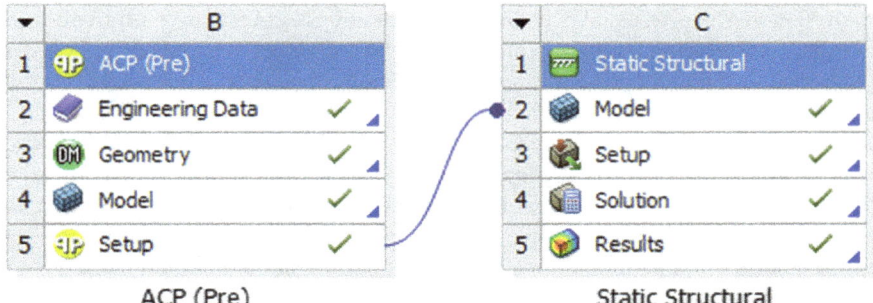

Fig. 8 Schematic view of static structural for hybrid laminate composite

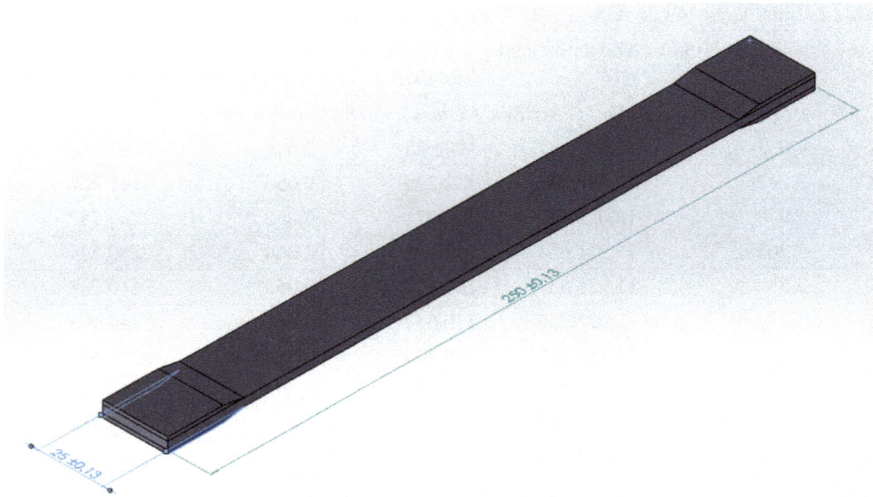

Fig. 9 Isometric view of tensile test sample using ASTM D3039 standard

Thickness of the sample initially was set 1 mm. The real thickness of the sample composite will be defined in the ANSYS Composite PrepPost model based on the composite layup. Then, assigned the type of material for the sample to be analysed. All the properties of chosen material were earlier defined in engineering data module. In this case study, a Glass and Kenaf fiber were assigned as a material of hybrid composite laminate.

3.2 Material Properties

Material properties is located in engineering data step is a compulsory information in analysis. By following a subsequent, the material properties must be defined earlier before modelled the sample in geometry. In this study, the material involved in hybrid laminate composite, E-Glass and Kenaf fibres, will be defined their properties in engineering data. The hybrid laminate E-Glass and Kenaf are categorized in orthotropic material which has three orthogonal symmetry planes. Therefore, a new material was created with new properties which obtained from the experimental. Inserted all the data such as density, orthotropic elasticity and also orthotropic stress and strain limit. The tabulation material properties data of each fiber is shown on Table 2.

Table 2 Material properties of E-Glass and Kenaf fibre

Properties	E-glass	Kenaf
Density (g/cm^3)	1.9	1.3
Exx (MPa)	48,750	13,919
Eyy (MPa)	12,000	22,800
Ezz (MPa)	12,000	22,800
Vxy	0.19	0.324
Vyz	0.31	–
Vxz	0.30	–
Gxy (MPa)	5500	2800
Gyz (MPa)	5000	2677
Gxz (MPa)	5500	2800

3.3 Generating the Mesh

Meshing is the most critical part of pre-processing in simulation. An effective mesh can give maximum accuracy result and reduce the computational time. For this case, the composite sample is rectangular simple plate, therefore, a quadrilateral with 245 elements for both GKG and KGK were used. The total number of nodes is 300. For the validation, a fine meshed was used since it will give an accurate results compared than coarse and medium. So, in order to get more accuracy result for this laminate composite sample, meshing should be finer and therefore mesh it with own parameter. The mesh element of the composite sample is depicted in Fig. 10.

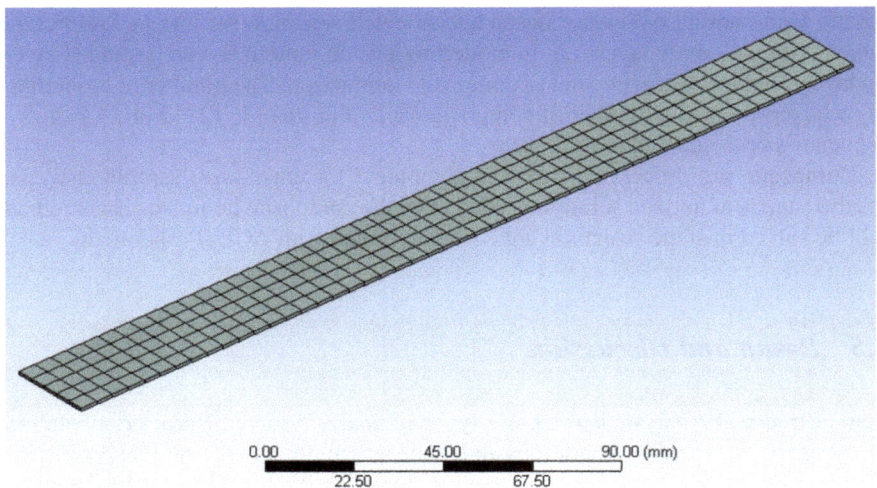

Fig. 10 Mesh generated with parameters set as per requirement

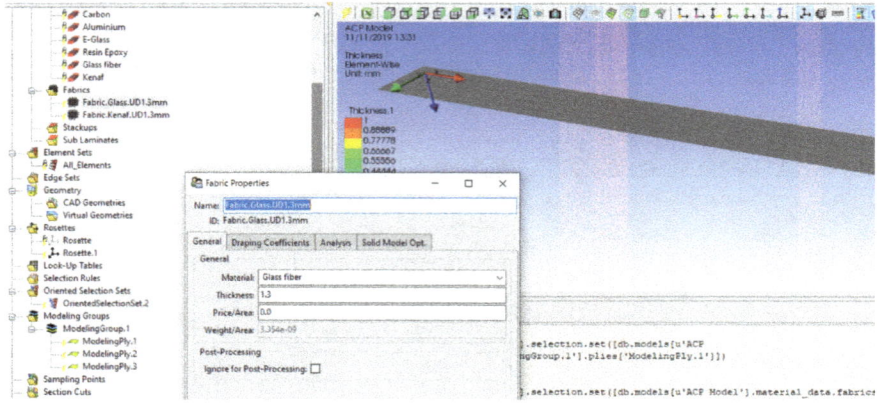

Fig. 11 Define a fabric used and the thickness

3.4 ACP PrepPost Setup

As it is known that, composite laminate has numerous layers of materials, different thicknesses and different ply orientation. In subsequent step, ANSYS Prepost (ACP) will be defined all that criteria such as a material used and its thickness, each lamina arrangement and fibre orientation. For this case study, there are two variations of composite lamination, Glass-Kenaf-Glass and Kenaf-Glass-Kenaf. Both of the sample thickness are 4 mm and have the ply orientation of 90°, 0° and 90°. A purpose of rosette is to define a fiber direction reference of 0° while oriented selection set is to define the direction of layup (Basri et al. 2019a). For this composite laminate, rosette with parallel type was chosen and oriented selection set was in Y-direction which is area to apply layers on. In modelling ply, all materials were arranged layer by layer similar to experimental of composite lamination. The number of modelling ply was represented a number of composite layer. Figures 11, 12, 13 and 14 shown sequence step contained in ACP setup.

Composite pre-processing part was completed for tensile test sample analysis. Further step is to find the solution needed from the composite laminate. This method will be solved in static structural analysis that linked with ACP (Pre) domain.

3.5 Result and Discussion

Static Structural Analysis is the basic types of analysis solver. Before run solution of the analysis, boundary conditions and loads was applied to the sample. In experimental of tensile test, the sample was pulled slowly until it breaks so the load applied in simulation should be similar to the experiment. This concept of stress will categorize as normal stress which is the force was applied in x-direction at one side while

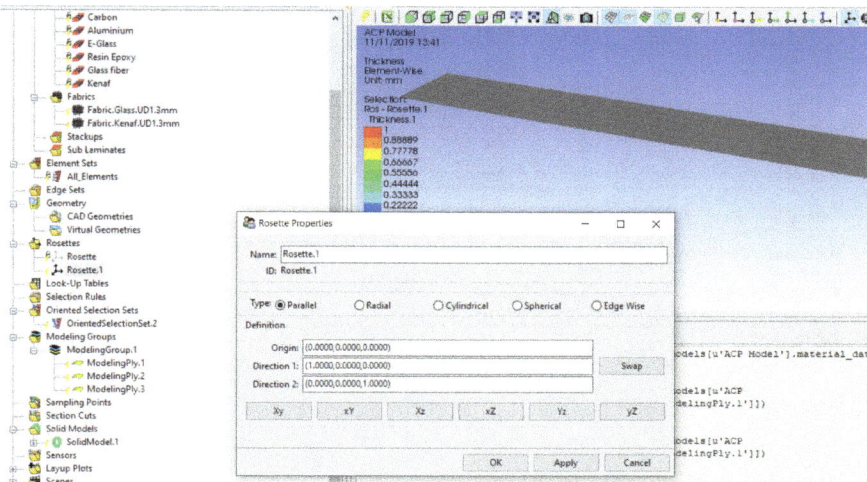

Fig. 12 Set the rosette of the sample

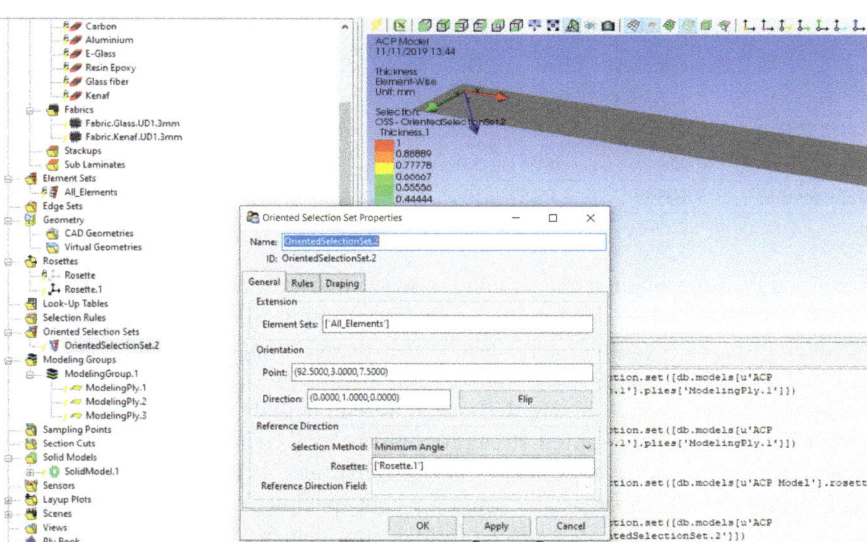

Fig. 13 Define the direction of oriented selection

boundary condition was put at another side in same direction. From the analysis of finite element and experimental for both specimen, result stress and strain (Table 3) are successful in validation range of 15% while percentage error for deformation result of both specimens achieved 100%. The simulation result is more accurate than experimental result. It will be supported with a several factors were occurred when

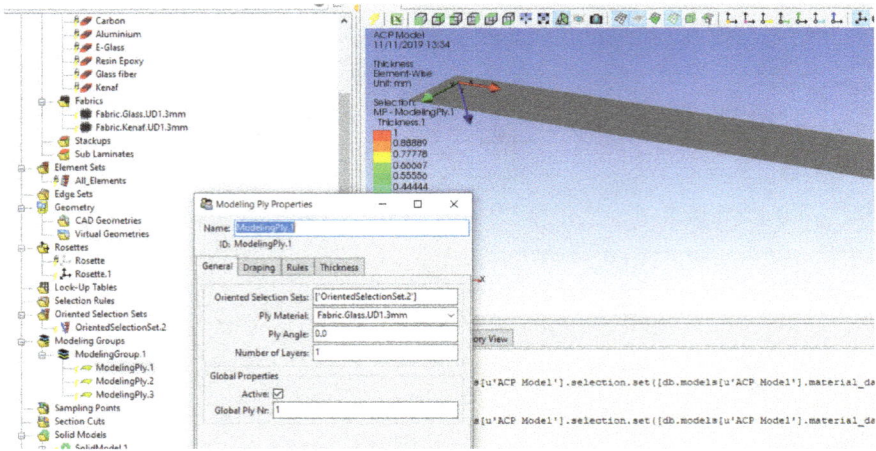

Fig. 14 Define a ply orientation and layers

Table 3 Comparison of analysis result from experiment and finite element analysis

	GKG			KGK		
	Experimental	FEA	Differences (%)	Experimental	FEA	Differences (%)
Stress (MPa)	106.64	120.89	13.4	70.17	79.787	13.7
Strain	0.054	0.057	5.6	0.042	0.036	14.3
Deformation	5.406	13.869	156	4.163	8.581	106

performed the experimental and simulation work. In experimental, it is possible that environmental or human error influence the result obtained such as the strength of specimen grip, a temperature applied in curing process, improper binding for each layer of fibre, etc. (Cook 1995). Compare to simulation process, it can be considered as a perfect process which is zero deformity. Based on Table 3, it clearly shows that composite of GKG have a high strength compare to KGK. Thus, in term of high strength and stiffness, a composite of GKG is most suitable for UAV wing skin. All finite element results of stress and strain for both specimens are shown in Figs. 15, 16, 17, 18, 19 and 20. The loads can get from the raw data of testing performed. Table 3 shown a comparison analysis result of experiment and finite element analysis for both hybrid laminate composite.

Glass-Kenaf-Glass Laminate Composites
See Figs. 15, 16 and 17.

Kenaf-Glass-Kenaf Laminate Composites
See Figs. 18, 19 and 20.

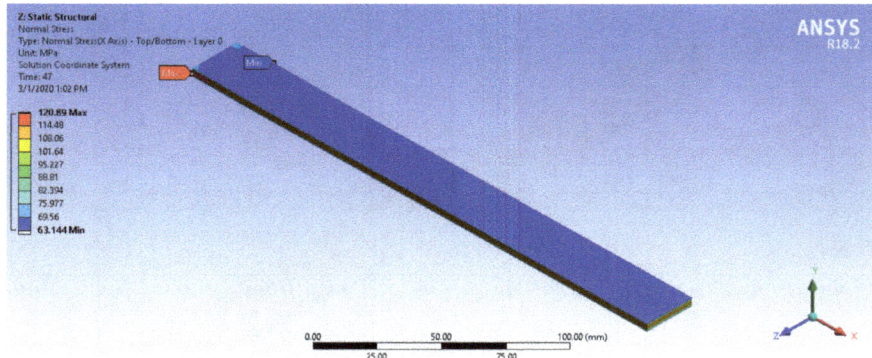

Fig. 15 Normal tensile stress of GKG

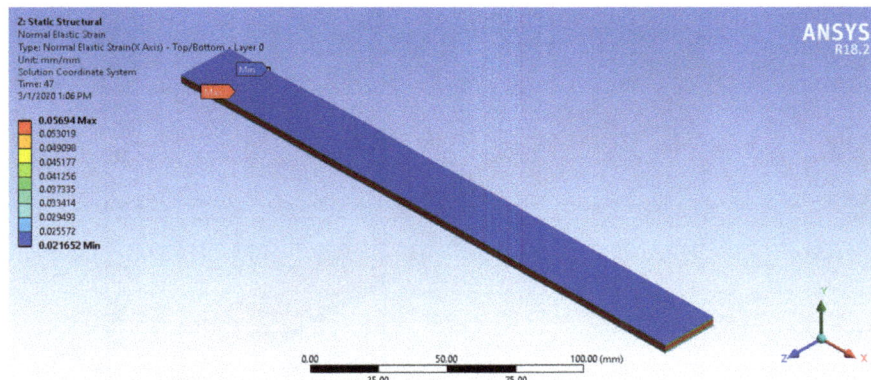

Fig. 16 Normal elastic strain of GKG

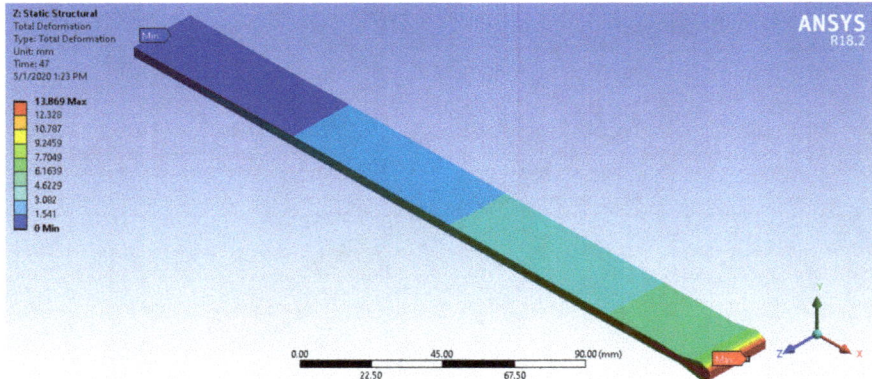

Fig. 17 Total deformation of GKG

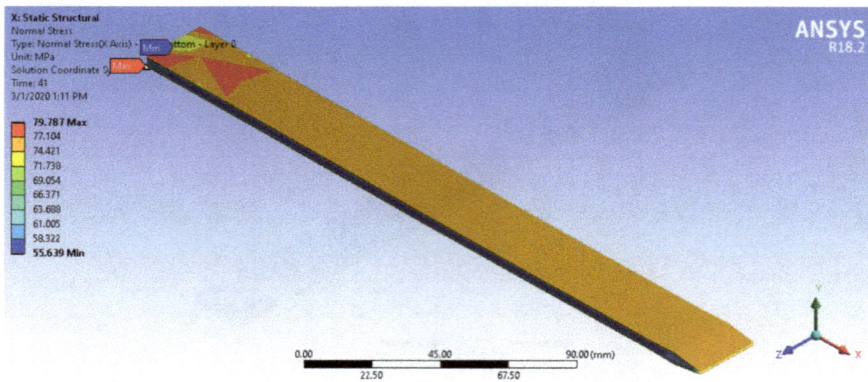

Fig. 18 Normal tensile stress of KGK

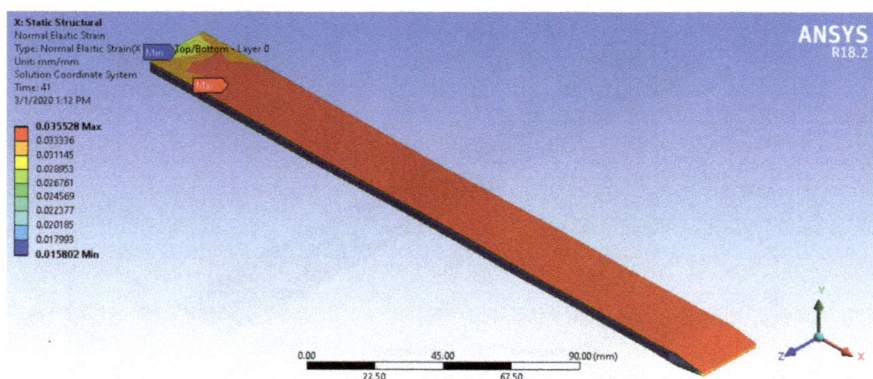

Fig. 19 Normal elastic strain of KGK

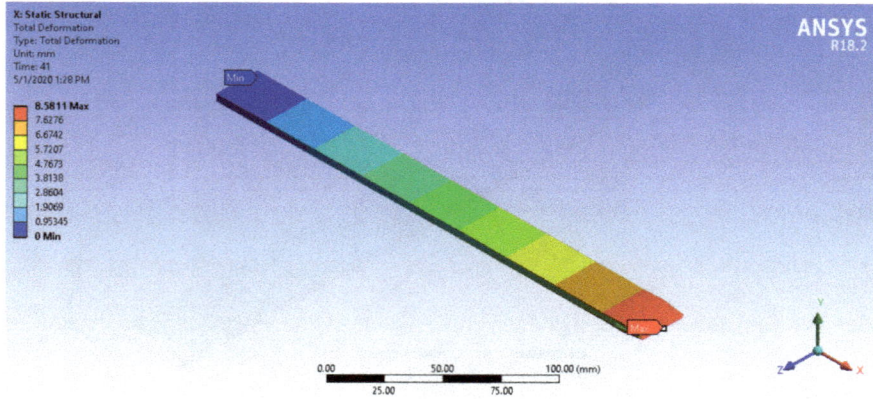

Fig. 20 Total deformation of KGK

4 Conclusion

In this study, it has been focused on engagement of natural composite in a hybrid laminate composite by fabricating two variations of GKG and KGK sample and validated the result obtained from experimental using finite element analysis. Both of the composites were compared in order to define which have the high strength to applied in UAV wing skin. From the finite element analysis, it was found that composite laminate of GKG have a maximum strength of 120.89 MPa compared to KGK which only have 79.787 MPa of maximum stress. In validation between same variation, it was showed a good agreement in stress and strain results in range of 5–15% differences but giving a percentage error for deformation due to the several factors was discussed before. Therefore, from the overall result was observed, it was showed that a hybrid natural-synthetic composite of glass-kenaf-glass giving better properties for applying this composite to a UAV wing skin.

ACKNOWLEDGMENTS The authors would like to thank Universiti Putra Malaysia for financial support through the Geran Putra Berimpak, GPB 9668200. The authors would like to thank the Department of Aerospace Engineering, Faculty of Engineering, Universiti Putra Malaysia and Laboratory of Biocomposite Technology, Institute of Tropical Forestry and Forest Product (INTROP), Universiti Putra Malaysia (HICOE) for the close collaboration in this research.

References

Al-Qrimli HF, Mahdi FA, Ismail FB (2015) Carbon/epoxy woven composite experimental and numerical simulation to predict tensile performance. Adv Mater Sci Appl 4:33–41. https://doi.org/10.5963/amsa0402001

Alger M (1997) Polymer science dictionary, 2nd edn. Springer Publishing

Basri EI, Mustapha F, Sultan MTH et al (2019) Conceptual design and simulation validation based finite element optimisation for tubercle leading edge composite wing of an unmanned aerial vehicle. J Mater Res Technol 8:4374–4386. https://doi.org/10.1016/j.jmrt.2019.07.049

Basri EI, Sultan MTH, Faizal M et al (2019) Performance analysis of composite ply orientation in aeronautical application of unmanned aerial vehicle (UAV) NACA4415 wing. J Mater Res Technol 8:3822–3834. https://doi.org/10.1016/j.jmrt.2019.06.044

Biswas S, Anurag J (2019) Fabrication of composite laminates. Reinf Polym Compos 39–53. https://doi.org/10.1002/9783527820979.ch3

Cook RD (1995) Finite element modeling for stress analysis. Wiley

Saravanan SK, Vetrivel R (2016) Experimental analysis of carbon/glass fiber reinforced epoxy hybrid composite with different carbon/glass fiber ratios. Int J Innov Res Sci Eng Technol (An ISO Certif Organization)

Prasad N, Ramachandran A (2017) Numerical analysis of hybrid carbon fiber composite specimen and validation of results. Int J Mech Eng Technol 8:75–88

Ramesh M, Nijanthan S (2016) Mechanical property analysis of kenaf-glass fibre reinforced polymer composites using finite element analysis. Bull Mater Sci 39:147–157. https://doi.org/10.1007/s12034-015-1129-z

Rao SS (2004) The finite element method in engineering, 4th edn. Elsevier, Butterworth-Heinemann

Reddy JN, (2003) Mechanics of laminated composite plates and shells theory and analysis, pp 840. https://doi.org/10.1007/978-1-4471-0095-9

Salleh Z, Yunus S, Masdek NRNM, et al (2018) Tensile and flexural test on kenaf hybrid composites. IOP Conf Ser Mater Sci Eng 328. https://doi.org/10.1088/1757-899X/328/1/012018

Numerical Approach to Evaluate Failure of Composites Under Ballistic Impact

Suhas Yeshwant Nayak, K. Rajath Shenoy, B. Satish Shenoy, M. T. H. Sultan, and Chandrakant R. Kini

Abstract This chapter presents a review of numerical approach to evaluate the strength and the failure of composites in a ballistic impact environment. The chapter is a compilation of various findings and theories which are utilized in estimating the performance of the composite system. The chapter is broadly classified into three categories viz. energy model which forms the fundamental numerical approach based on law of energy conservation, strength and failure models used to assess the performance and Finite Element Method approach which features various stages to formulate and solve a composite material interaction system for ballistic tests using a wholesome software. To address the inherent diversity of the composite systems, a plethora of models and theories have been discussed. The results obtained from these numerical models are generally comparable to the experimental results and thus, the future shows immense scope for emergence of better numerical models which can imitate real life performance with more accuracy.

Keywords Numerical models · Finite element methods · Ballistic impact · Composite materials

S. Y. Nayak · K. R. Shenoy
Department of Mechanical and Manufacturing Engineering, Manipal Institute of Technology, Manipal Academy of Higher Education, Manipal, Karnataka 576104, India

B. S. Shenoy (✉) · C. R. Kini
Department of Aeronautical and Automobile Engineering, Manipal Institute of Technology, Manipal Academy of Higher Education, Manipal, Karnataka 576104, India
e-mail: satish.shenoy@manipal.edu

M. T. H. Sultan
Department of Aerospace Engineering, Faculty of Engineering, Universiti Putra Malaysia, 43400 UPM Serdang, Selangor, Malaysia

© Springer Nature Singapore Pte Ltd. 2021
M. T. H. Sultan et al. (eds.), *Impact Studies of Composite Materials*, Composites Science and Technology, https://doi.org/10.1007/978-981-16-1323-4_11

1 Introduction

Composites have been a preferred material system for ballistic application for years (Salman et al. 2015). There has been extensive research going on in improving the properties. Conventional impact resistant materials are getting replaced by composites mainly due to lesser cost, lighter weight and better performance, especially in the field of body armour, ground and aerial vehicles (Karahan et al. 2015; Mostafa et al. 2016; Zakikhani et al. 2016). Ballistic testing are generally categorized into either low velocity or high velocity testing, depending upon the velocity of impact of the projectile (Ismail et al. 2019).

The testing of the composite properties is done experimentally or analytically. Experimental procedure requires preparation of required samples and subjecting them to projectile shot from a ballistic equipment. The tested samples are then checked for its performance by assessing its ballistic limit and its suitability for providing protection as per different levels set by standards are established (Zulkifli et al. 2019). The analytical process involves implementation of various numerical models on the composite panel taking its properties into consideration and solving the equations of system interaction either by taking the system in its entirety or with finite element techniques. The performance is then evaluated for stress distribution, delamination failures, penetration based failure and the ballistic limit of the composite panel under consideration based on the numerical results obtained either in visual model form, tabular or graphical representation (Ansari et al. 2017).

The experimental testing methodology is usually associated with a lot of drawbacks. The standards requires multiple samples to be tested for every parameter changed in the experiment, which adds up to a lot of specimens to be produced. These specimens and the tests conducted on them costs a lot of time and money. The damage evaluation either by destructive or non-destructive testing also requires suitable setups and money for these tests. Skill to create specimens without any faults or voids in them is challenging, especially in composites which are fabricated using hand layup technique. Damage study often necessitates a skilled researcher who has ample amount of knowledge in the field.

Analytical approach makes it easier in these regards. It is often used by many researchers or engineers in the companies to predict the behaviour by creating a numerical or a virtual material system. This is usually done using a computer for the ease of calculation. The results are highly reproducible and gives a near ideal prediction of the system interaction. It is very easy to use a single material model and change the testing parameters any number of time to get desired results without any expense of materials or cost generally. One needs to have sufficient knowledge about choices of numerical models to be implemented, necessary software or similar system for calculation and data regarding the properties of the composite panel, projectile or any other surrounding attributes. Some data may be necessary to be obtained through some experimental processes, but in most cases these are not as frantic as full scale experimental procedure.

As discussed earlier, one of the main prerequisite for using analytical approach towards ballistic testing is having a deep knowledge about the numerical models and finite element methods implemented. Almost all failures and setbacks are often due to improper or misinterpretation of these concepts which leads to erroneous or misleading results. Thus the current chapter points out various numerical models and criterions being implemented under the necessary conditions, which are at the forefront of the research in the current days. The chapter shall also deal with certain important guidelines regarding Finite Element Method (FEM) techniques being very specific in scenarios of ballistic testing on composites, thereby making sure that the reader is well informed and makes the best use of this method.

2 Energy Model

The energy model is one of the basic and widely used model in numerical analysis of ballistic response of the composite system. It is based on the law of conservation of energy of the projectile—composite barrier system (Langston 2017). The total energy lost by the projectile during its penetration into the composite barrier is utilised or dissipated through various modes of failures in the composite. Some of the basic assumptions during the application of this model is as follows (Langston 2017):

- The composite system is stationary and the projectile is moving with uniform velocity at time t = 0 s
- Composite plate is flat
- The composite panel is either at $0°$ or $90°$ with respect to horizontal
- The angle of incidence of projectile is $90°$
- No mass loss from projectile
- The projectile is cylindrical in shape
- The projectile motion is constant across each discrete time interval.

Delamination of the layers can occur through three different types of modes, represented as modes I, II and III. In a ballistic impact environment, modes I and II are more prominent and have more contribution in the damage mechanics compared to the third mode. Figure 1 shows the three modes of failure of the layers.

Mode I refers to the propagation of crack between the layers when the tensile load is applied perpendicular to the plane of crack growth while mode II has the load applied linearly in a plane parallel to the crack plane. Mode I is also referred as opening mode and mode II as sliding mode based on the movement of the layers (Mallick 2007).

When either of the modes of failure occur, strain energy release rate is calculated to study the overall integrity of the body after the damage has been made. The implication of these strain energy release rates are explained later in the chapter.

Failure due to transverse cracking is more prominent in carbon/ epoxy or glass/epoxy composite system but studies have found no role of it in failure of certain systems like that of Ultra High Molecular Weight Polyethylene (UHMWPE)

(a)

(b)

(c)

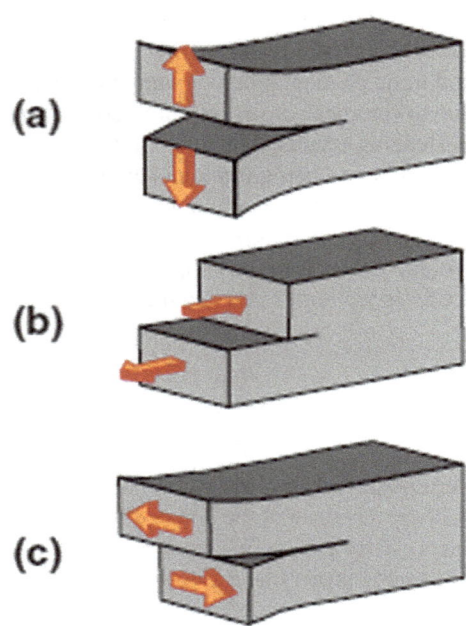

composites. Layer failure is observed when the instantaneous compressive tensile strain and shear stresses is exceeds the maximum value (Langston 2017).

The compressive stress is given by (Langston 2017):

$$\sigma_{c,i} = \frac{\sigma_{t,i}}{\gamma}\left(\frac{Ef + Em}{Em}\right) \tag{1}$$

where σc = compressive stress in the ply and σt = tensile stress in the ply. E is the young's modulus of the fiber and matrix and γ is the possion's ratio. The tensile strain due to the compression in the ply is further given by the relation (Langston 2017):

$$\varepsilon_{c,i} = \frac{\sigma_{c,i}}{Em}\left(1 - \frac{\gamma^2 Ef}{(Ef + Em)}\right) \tag{2}$$

This strain is commonly added to the maximum tensile strain due to cone formation (mentioned in the upcoming subsection) to get a better estimate of energy absorption. Nonetheless, the contribution of tensile strain effect due to this compressive loading on the composite by itself alone is generally neglected since its contribution in energy absorption is very small as compared to the energy absorption due to cone formation (Langston 2017).

The energy model equation is given as (Langston 2017):

$$E_{total} = E_{KE\ projectile} + E_{tensile} + E_{delamination} + E_{acceleration} + E_{cracking} + E_{shear}$$
$$(3)$$

Here, E_{total} is the initial kinetic energy of the projectile and $E_{KE\ projectile}$ is the residual kinetic energy of the projectile exiting the composite. The remaining terms of this equation is explained in detail in the upcoming subsections. Each of these terms show unique way of energy absorption mechanics taking place in the composite specimen subjected to the ballistic test.

2.1 Tensile Energy Loss Due to Cone Formation

Just before the projectile completely penetrates through the ply, composite deforms under the force. Three types of deformation waves sweep across the surface of the composite surface perpendicular to the incidence of the projectile—elastic wave, plastic wave (also called strain wavelets) and transverse wave as seen in Fig. 2. The elastic wave travels the fastest followed by the plastic wave and then transverse wave. The plastic wave causes the material to flow inwards towards the impact point. Once the plastic waves sweep, the tensile strain of the material does not change but the material moves in transverse direction, parallel to projectile.

The velocity of plastic wave upon which the strain relies is dependent upon the instantaneous slope of stress–strain curve at yield limit and density of the composite (Langston 2017).

$$C_p = \sqrt{\frac{1}{\rho}\left(\frac{d\sigma}{d\varepsilon}\right)_{\varepsilon=\varepsilon_p}}$$
$$(4)$$

The transverse wave velocity is given by (Langston 2017):

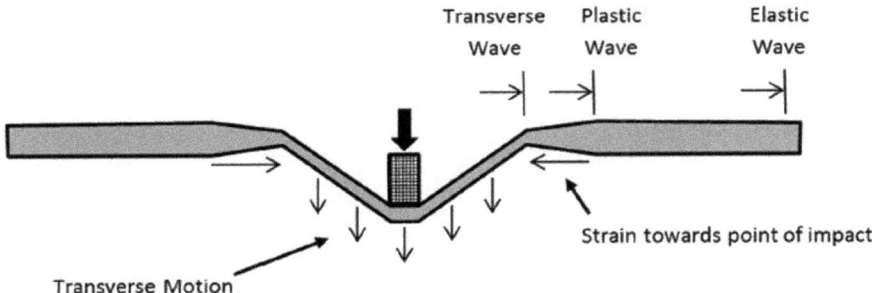

Fig. 2 Three types of wave sweeping across the composite (Langston 2017)

$$C_t = C_p\sqrt{\frac{\varepsilon_p}{1+\varepsilon_p}} \tag{5}$$

Due to the anisotropic property of the composites, it is evident that the strain energy too is related the orientation of the fibers in the composite. On a superlative stand, the total strain energy in an anisotropic material model is associated with individual strain energies along 6 eigen vectors. In an isotropic system, the radial damage will be uniform as the vectors are equal but in anisotropic model, the individual eigen vector based strain energies have different values and limits to determine the extent of failure (Biegler and Mehrabadi 1995). For a composite material system, these 6 eigen vector can be reduced to fewer numbers based on the material properties along the concerned direction. This can be done to narrow down on similar and dissimilar energy eigen mode and fix values to estimate damage in each eigen vector.

A simplified version of the above methodology is with implementation of Krenchel composite efficiency factor. The anisotropy of the composite is quantified as a single factor β_θ which depends on the proportion of each fiber orientation of the total fiber content a_n and the fiber orientation angle θ_f (Langston 2017).

$$\beta_\theta = \sum a_n cos^4\theta_f \tag{6}$$

This efficiency factor is then used for formulating the Young's modulus of the composite by the modified rule of mixture equation (Langston 2017):

$$E_c = \beta_\theta E_f V_f + E_m\left(1 - V_f\right) \tag{7}$$

where E_f and E_m are the modulus of the fiber and matrix respectively and V_f is the fiber volume fraction. This method gives a very good approximation by quantifying the anisotropy and giving an overall energy absorption value, especially for composites having high volume fraction of high stiffness fibres (Langston 2017).

The tensile strain energy due to cone formation can be found out by knowing the strained volume V (excluding the failed layers due to tensile and shear stress) and the composite modulus E_c = obtained from the rule of mixture (Langston 2017).

$$E_{tensile} = \int_0^\varepsilon (E_c\varepsilon)d\varepsilon.V \tag{8}$$

2.2 Energy Absorbed During Delamination Between the Layers

The strain energy release rate due to delamination is assumed to be following a mixed model I and II of shear failure. The strain energy release rate is the rate at which the potential energy of the body is lost due to application of load. This energy is responsible for maintaining the integrity of the composite. The energy absorption due to delamination of layers is dependent upon the crack area and this strain energy release rate (Langston 2017).

$$E_{delamination} = \sum \pi A_{ql} G_c N_{IF,i} r_{T,i}^2 P_{IF} \tag{9}$$

where G_c is the combined critical energy release rate, N_{IF} is the number of remaining layers of composite which has not yet failed, A_{ql} is the quasi lemniscate reduction factor, r_T is the radius of failure zone and P_{IF} is a parameter which denotes a fraction of layers which do not undergo delamination (generally assumed to be around 0.08).

2.3 Energy Absorption During Composite Mass Acceleration

When the projectile strikes the composite surface, it undergoes a conical deformation and the deformed mass actually moves in the direction of the projectile for a very short interval of time. Energy is absorbed due to this sudden acceleration and deceleration of the composite mass. It should be noted that this energy absorbed is concerned only with intact layers and once a layer has failed and delaminated from the rest of the mass, it no longer has any acceleration (Langston 2017).

$$\begin{aligned} E_{acceleration} &= \sum \frac{mass \times velocity^2}{2} \\ &= 0.5 A_{ql} \pi \sum \left(r_{T,i}^2 - r_{T,i-1}^2\right).\left(h_c - N_{SP,i} h_L - N_{T,i} h_L\right) \rho_c v_i^2 \end{aligned} \tag{10}$$

Here h_c represents the thickness of entire composite panel, h_L is the thickness of one layer, N_{SP} and N_T denotes the number of layers failed by shear or tensile failure, ρ_c is the composite density, v is the velocity of the layer and r_T is the damage cone radius.

Equation 8 holds good for damage radius $r_T < L_T/2$ where L_T is the length of the side of square specimen. If the damage area exceeds the above value, then the value of the same must be found using trigonometric relations.

2.4   Energy Loss Due to Shear Plugging

Shear stresses are developed around the perimeter of the projectile as it hits the composite panel. This stress due to plugging can be given as (Langston 2017):

$$\sigma_{SP,i} = \frac{F_{P,i}}{\pi d_P \left(h_c - N_{SP,i} h_L - N_{T,i} h_L \right)} \tag{11}$$

where d_P is the diameter of projectile and F_P is the contact force which is expressed as (Langston 2017):

$$F_{P,i} = \left[m_P + \rho_c (N_T + N_{SP}) h_L \left(\frac{\pi d_P^2}{4} \right) \right] dC_i \tag{12}$$

where m_P is the mass of projectile and dC_i is the deceleration of projectile. If the composite through thickness strength $S_{SP} > \sigma_{SP}$, then (Langston 2017):

$$E_{shear} = \sum S_{SP} \pi d_P S_{P,i}^2 \tag{13}$$

2.5   Energy Absorption Due to Matrix Cracking

During the failure of the composite, the matrix undoubtedly undergoes cracking. But it should be noted that only a percentage of total matrix P_m cracks by mode I of strain energy release rate G_I. N_L is the number of intact layers.This cracking is assumed to be only inside the intact portion of the composite (Langston 2017).

$$E_{cracking} = \sum 2\pi P_m A_{ql} r_{P,i} G_I N_{L,i} h_L \tag{14}$$

Langston (2017) used the energy model in his research regarding evaluation of ballistic performance of UHMWPE and compared it with experimental results. Both the results showed high correlation with almost similar failure modes and ballistic properties. Pach et al. (2017) implemented this model during the numerical investigation of properties of a hybrid composite shield made of SBS and metal plate laminates such as TiC, Al_2O_3 and steel. It was concluded that the energy distribution and absorption values from numerical analysis were in agreement with the experimental values (Pach et al. 2017).

All these subsections (2.1–2.5) are the major contributors for the energy transfer and absorption in the model. During the projectile-panel interaction, all these independent energy absorption mechanics happen simultaneously. Each of these mechanisms are assumed to take place based on the system properties. Langston used Eq. 3 as the fundamental equation in his work and used all the terms which was observed crucial for investigating the ballistic properties of UHMWPE composites. Equation 3 is not universal and has other energy components. These energy components are either material specific or have their significance ascertained based on the presence or absence of other modes of energy absorption.

Pasquali and Gaudenzi (2017) in their research on effect of curvature on thin fabric composites implemented new terms such as energy absorption due to breaking of primary yarns and energy absorption due to bending of primary and secondary yarns along with other terms of Eq. 3. Balaganesan et al. (2014), while investigating the energy absorption mechanics in nanocomposites laminates during ballistic tests, made similar changes to Eq. 3. The modified equation had new terms and energy absorption mechanics which included energy absorbed due to tensile failure of the fibers and deformation of secondary of the fibers along with other terms. Thus, it is evident that Eq. 3 is generally modified to accustom the energy absorption mechanics based on the material properties and system interaction.

3 Strength and Failure Models

Apart from energy model, several other numerical models are utilized for determining the strength and failure of the composite. The choice and application of these models are based on the type of material system, testing conditions, possible modes of failure and other parameters. These models (Cao et al. 2015) are utilized in finite element methodology (discussed further on in this chapter) to solve at each node point of the structure. A brief description of each model is given below.

3.1 Johnson—Cook Model (JC Model)

Proposed by Gordon R. Johnson and William H. Cook in 1985, this model is used to characterize the material performance which is subjected to high strain rates and thermal shock load (rising due to sudden impact of the projectile onto the panel surface). This model is generally utilized on performance check of ductile material system in ballistic test scenario such as composites made or aluminium layers, titanium layers, super alloys laminae or metal–matrix system.

This model actually constitutes of two models—strength and failure predicting model. The material's yield stress is found out by strength model (Cao et al. 2015).

$$\sigma_y = \left(A + B\varepsilon_p^n\right)\left(1 + Cln\dot{\epsilon}^*\right)\left(1 - T^{*m}\right) \tag{15}$$

where, A, B, n, m and C are material constants, $\dot{\epsilon}^*$ is the dimensionless plastic strain rate, ε p is the equivalent plastic strain. T* is the homologous temperature given by (Cao et al. 2015):

$$T^* = \frac{T - T_R}{T - T_M} \tag{16}$$

where T_M and T_R and melting and room temperatures respectively.

Constant A is the yield strength of the material when the stress rate is low at a given temperature (PNAE G-7–002-86 1989). The m parameter denotes the thermal softening of the material. These values can be obtained from the material data sheet at different temperatures.

The B and n constants take the isotropic (static) strengthening during strain into consideration. These parameters are determined based on data on the material behavior during low stressing rate (quasistatic) (Sobolev and Radchenko 2016). The parameter C is responsible for kinematic strengthening due to strain effect which again is obtain from the material data sheet.

In the failure model, the damage parameter D ($0 \leq D \leq 1$) represents the extent of damage in the system. If $D = 1$, failure is observed (Cao et al. 2015).

$$D = \sum \Delta\varepsilon / \varepsilon^f \tag{17}$$

$\Delta\varepsilon$ is the equivalent plastic strain increment in one time step and ε^f is the material fracture strain which is obtained from damage parameters (D1, D2, D3, D4, D5) through Hopkinson pressure bar test (Cao et al. 2015).

$$\varepsilon^f = \left[D_1 + D_2 exp\left(D_3\sigma^*\right)\right]\left(1 + D_4 ln\dot{\varepsilon}^*\right)\left(1 + D_5 T^*\right) \tag{18}$$

Work done by Holmen et al. (2017) pointed out that the JC model predicted the failure mode, especially the shape of the failed surface very accurately upon comparison with experimental tests.

3.2 Johnson—Holmquist Material Damage Model (JH-2 Model)

This model is quite similar to JC model except for the fact that this model is used to simulate mechanical response of brittle materials like ceramics undergoing high strain rate and large deformations as seen in case of ballistic tests. The damage parameter D is the same as seen in JC model. The strength of the material at any given point is given by (Cao et al. 2015):

$$\sigma = \sigma_i - D\left(\sigma_i - \sigma_f\right) for \varepsilon D \in (0, 1) \tag{19}$$

where σ_i is the initial material strength ($D = 0$) and σ_f is fracture strength ($D = 1$) given by (Cao et al. 2015):

$$\sigma_i = A\left(p^* + T^*\right)^N (1 + C ln \dot{\epsilon})$$ (20)

$$\sigma_f = B p^{*m} (1 + C ln \dot{\epsilon})$$ (21)

where A, B, C, N and m are constants, p* and T* are current and maximum hydrostatic pressures normals at Hugoniot elastic limit.

Cao et al. (2015) used both JC as well as JH-2 models while dealing with a composite—projectile system having both ductile Ti phase and brittle natured Al_3Ti phase in the composites for strength and failure check. The dynamic response of the phases were as per these models and helped in establishing the multiple energy absorption mechanisms (Cao et al. 2015).

3.3 Tsai—Wu Failure Criterion

Tsai—Wu model is used to determine the failure of anisotropic composite materials which have varied compressive and tensile strengths. The failure parameter F = 1 indicates the failure of model (Tsai and Wu 1971).

$$F = \sum F_i \sigma_i + F_{i,j} \sigma_i \sigma_j \leq 1$$ (22)

For an orthotropic material having symmetric planes in 3D space, the equation simplifies as (Tsai and Wu 1971):

$$F_1\sigma_1 + F_2\sigma_2 + F_3\sigma_3 + F_{11}\sigma_1^2 + F_{22}\sigma_2^2 + F_{33}\sigma_3^2 + F_{44}\sigma_4^2 + F_{55}\sigma_5^2 + F_{66}\sigma_6^2$$
$$+ 2F_{12}\sigma_1\sigma_2 + 2F_{13}\sigma_1\sigma_3 + 2F_{23}\sigma_2\sigma_3 \leq 1$$ (23)

If σ_{iT} and σ_{iC} represent uniaxial tensile and compressive stresses, τ_{12}. τ_{13}, τ_{23} represent shear strengths, the parameters for F_i and F_{ij} is given as follows (Tsai and Wu 1971):

$$F_i = \frac{1}{\sigma_{iT}} - \frac{1}{\sigma_{iC}} \; for \; i = 1, 2, 3; \; else \; F_i = 0$$ (24)

$$F_{ii} = \frac{1}{\sigma_{iT}\sigma_{iC}} \; for \; i = 1, 2, 3$$ (25)

$$F_{44} = \frac{1}{\tau_{23}^2}, F_{55} = \frac{1}{\tau_{31}^2}, F_{66} = \frac{1}{\tau_{12}^2}$$ (26)

Let the failure strengths in equibiaxial tensions be (Tsai and Wu 1971):

$$\sigma_1 = \sigma_2 = \sigma_{b12}, \sigma_1 = \sigma_3 = \sigma_{b13}, \sigma_2 = \sigma_3 = \sigma_{b23}$$ (27)

then:

$$F_{mn} = \frac{1}{2\sigma_{bmn}^2}\left[1 - \sigma_{bmn}(F_m + F_n) - \sigma_{bmn}^2(F_{mm} + F_{nn})\right] \text{ for } mn = 12, 13, 23$$

(for each m and n, separate values are assigned)

(28)

Nayak et al. (2017) used material law 14 of Hyperworks, which implements this model for strain rate effects and damage evaluation of aramid fiber and epoxy based composite being tested for ballistic properties at various projectile impact velocities.

3.4 Chang–Chang Failure Model

This failure model, originally formulated by Chang and Chang (1987), takes into consideration of the transverse strength along with longitudinal strength of the composite while proposing the failure modes of fiber and matrix separately. Failure is evident when any of the equation equals 1 (Altair Engineering 2020).

Failure criterion for fiber failure is as follows:

Tensile fiber mode (Altair Engineering 2020):

$$e_f^2 = \left(\frac{\sigma_{11}}{S_1}\right)^2 + \beta\left(\frac{\sigma_{12}}{S_{12}}\right)^2 \quad for \ \sigma_{11} > 0$$

(29)

Compressive fiber mode (Altair Engineering 2020):

$$e_c^2 = \left(\frac{\sigma_{11}}{C_1}\right)^2 \quad for \ \sigma_{11} < 0$$

(30)

Failure criterion for matrix cracking is as follows:

Tensile matrix mode (Altair Engineering 2020):

$$e_m^2 = \left(\frac{\sigma_{22}}{S_2}\right)^2 + \beta\left(\frac{\sigma_{12}}{S_{12}}\right)^2 \quad for \ \sigma_{22} > 0$$

(31)

Compressive matrix mode (Altair Engineering 2020):

$$e_d^2 = \left(\frac{\sigma_{22}}{2S_{12}}\right)^2 + \left[\left(\frac{C_2}{2S_{12}}\right)^2 - 1\right]\frac{\sigma_{22}}{C_2} + \left(\frac{\sigma_{12}}{S_{12}}\right)^2 \quad for \ \sigma_{11} < 0$$

(32)

where S_1 is the longitudinal tensile strength, S_2 is the transverse tensile strength, S_{12} is the shear strength, C_1 is the longitudinal compressive strength, C_2 is the transverse compressive strength and β is the shear scaling factor.

Material model 22 of LS-DYNA solver implements this model and upon application of model by by Berk et al. (2017) From a finite element model of hybrid carbon/aramid epoxy composite, researchers were able to illustrate the changes in the hole diameters on either surfaces of the panel for various velocities along with other parameters (Berk et al. 2017).

3.5 Hashin's Failure Criterion

The current failure criteria is considered only for unidirectional fiber composite. The failure modes are tensile/compressive of fiber and matrix are as shown (Hashin 1980):
Tensile fiber failure mode (Hashin 1980)

$$\left(\frac{\sigma_{11}}{\sigma_f^+}\right)^2 + \frac{\left(\sigma_{12}^2 + \sigma_{13}^2\right)}{\tau_f^2} \leq 1 \, for \, \sigma_{11} > 0 \tag{33}$$

Compressive fiber failure mode (Hashin 1980):

$$\left|\frac{\sigma_{11}}{\sigma_f^-}\right| \leq 1 \, for \, \sigma_{11} < 0 \tag{34}$$

Tensile matrix failure mode (Hashin 1980):

$$\frac{(\sigma_{22} + \sigma_{33})^2}{\sigma_T^{+2}} + \frac{\left(\sigma_{23}^2 - \sigma_{22}\sigma_{33}\right)}{\tau_T^2} + \frac{\left(\sigma_{12}^2 + \sigma_{13}^2\right)}{\tau_f^2} \leq 1 \, for \, \sigma_{22} + \sigma_{33} > 0 \tag{35}$$

Compressive matrix failure mode (Hashin 1980):

$$\frac{(\sigma_{22} + \sigma_{33})}{\sigma_T^-}\left[\left(\frac{\sigma_T^-}{2\tau_T}\right)^2 - 1\right] + \frac{(\sigma_{22} + \sigma_{33})^2}{4\tau_T^2} + \frac{\left(\sigma_{23}^2 - \sigma_{22}\sigma_{33}\right)}{\tau_T^2} + \frac{\left(\sigma_{12}^2 + \sigma_{13}^2\right)}{\tau_f^2} \leq 1$$
$$for \, \sigma_{22} + \sigma_{33} < 0 \tag{36}$$

Upon simplification by applying plane stress failure condition, the above modes become almost identical to Chang–Chang model.

Ansari et al. (2017) used this model on a unidirectional GRPF composite for the evaluation of its performance on oblique impact of the projectiles on it (Ansari and Chakrabarti, 2017a).

3.6 Elastic Continuum Damage Mechanics Model

This is a simple matrix model which gives the stress–strain relationship in a bidirectional orthogonal composite system wherein the stress vectors are oriented in the
fiber direction (Phadnis et al. 2013).

$$\begin{bmatrix} \varepsilon_{11} \\ \varepsilon_{22} \\ \varepsilon_{12}^{el} \end{bmatrix} = \begin{bmatrix} \frac{1}{(1-d_1)E_1} & -\frac{\gamma_{12}}{E_1} & 0 \\ -\frac{\gamma_{21}}{E_2} & \frac{1}{(1-d_2)E_2} & 0 \\ 0 & 0 & \frac{1}{(1-d_{12})2G_{12}} \end{bmatrix} \begin{bmatrix} \sigma_{11} \\ \sigma_{22} \\ \sigma_{12}^{el} \end{bmatrix} \tag{37}$$

where $\varepsilon = (\varepsilon_{11}, \varepsilon_{22}, \varepsilon_{12}^{el})$ is the strain vector, $\sigma = (\sigma_{11}, \sigma_{22}, \sigma_{12}^{el})$ is the stress vector, E_1
and E_2 are the young's moduli, G_{12} is the in-plane shear modulus, γ_{12} and γ_{21} are the
poisson's ratio. The damage parameters d_1 and d_2 are associated with fiber breakage
in the principal orthotropic directions and d_{12} is due to the matrix cracking because
of in-plane shear stress. Only the elastic shear response (denoted by superscript 'e_l'
in strain and stress matrix component) is being studied in this model. The plastic
behaviour is dealt in the next model.

3.7 Elastic–Plastic Shear Model

The in-plane shear response of composite system during a ballistic test is a culmina-
tion of non-linear behaviour due to both elastic, plastic and matrix micro-cracking.
The permanent deformation of the ply is generally due the matrix inelasticity due to
extensive cracking leading to plasticity. This is portrayed in this model which incul-
cates the hardening law and an elastic domain function F given by (Phadnis et al.
2013):

$$F = |\tilde{\sigma}_{12}| - \tilde{\sigma}_0(\bar{\varepsilon}^{pl}) \leq 0 \tag{38}$$

Equation 38 can be utilized in the hardening law equation (Phadnis et al. 2013),

$$\tilde{\sigma}_0(\bar{\varepsilon}^{pl}) = \tilde{\sigma}_{y0} + C(\bar{\varepsilon}^{pl})^p \tag{39}$$

where C is the initial effective shear yield stress, C and p are coefficients, $\bar{\varepsilon}^{pl}$ is
equivalent plastic strain due to shear deformation.

Phadnis et al. (2013) used both elastic continuum damage mechanics model and
elastic–plastic shear model in the finite element model of plain weave E-glass epoxy
composite to study the relation of thickness of composite on its ballistic limit.

3.8   Fracture Mechanics Model

According to this model, the ballistic performance of the composite is correlated to the quantitative measurement of crack density and damage tolerance of a crack. The impact energy is absorbed by the composite as the crack forms and propagates. The crack density ρ is measured by the formula (Cao et al. 2015),

$$\rho = \frac{l}{a} \tag{40}$$

where l is the total length of cracks and is the area under observation. The crack density gives us a qualitative estimate as to how much of energy is absorbed in the given area (Cao et al. 2015). This equation (Eq. 40) is not used for direct quantitative analysis of a 3D material models such as composite specimen. It is popularly used in the post processing activities of image analysing setups. The 2D image slices of a 3D specimen is taken using suitable techniques and the qualitative damage estimate is carried out.

The model is also used to predict the threshold load needed to delaminate an interface in a composite body. A double cantilever beam model is used to simplify the strategy of predicting the delamination as shown in the Fig. 3 (Abrate et al. 2015). The beam has a crack of length 2a already existing within it at a depth of h from either surface. P is the load acting on the central part of the beam.

The deflection of undamaged central portion is given by (Abrate et al. 2015),

$$\delta_u = \frac{3Pa^2(1 - \gamma^2)}{4\pi E(2h)^3} \tag{41}$$

The deflection of central portion of the beam having a crack of length 2a is given by (Abrate et al. 2015),

$$\delta_d = \frac{3Pa^2(1 - \gamma^2)}{8\pi Eh^3} \tag{42}$$

Fig. 3 Schematic of a double cantilever beam model (Abrate et al. 2015)

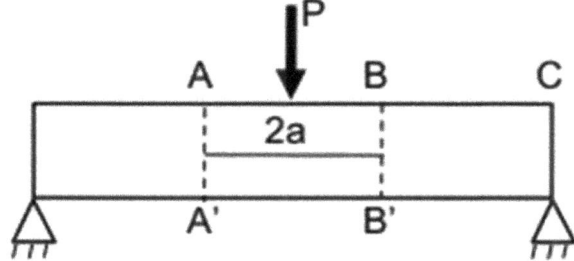

The composite delaminates due to mode II of shear failure model. Thus, the energy available to drive the crack to complete failure is given by (Abrate et al. 2015)

$$U = \frac{P(\delta_d - \delta_u)}{2} = \frac{9P^2a^2(1 - \gamma^2)}{64\pi \, Eh^3} \tag{43}$$

Therefore the delamination threshold load (DTL) is obtain by the equation (Abrate et al. 2015):

$$P_d^2 = \frac{64\pi \, Eh^3}{9a^2(1 - \gamma^2)}G_{IIc} \tag{44}$$

where G_{IIc} is the mode II critical strain energy release rate.

4 Finite Element Methodology

Finite element technique is often used along with the numerical model/ theories to solve a complicated material system interaction. The material system is divided into several tiny components called elements joined to each other at node points. The numerical model equations are solved for each of these node points with the appropriate boundary conditions to get desired results. This technique is used by several computer software that can solve millions of such nodal equations extremely fast, as compared to human capability.

Ballistic testing of composites using FEM is quite a challenge. Composites, on their own, are a very complicated system to model and simulate due to their intricate architecture and anisotropic (orthotropic) properties. The simulation requires specific solvers which can handle dynamic loading conditions. Sometimes experimental base data is required to fulfil boundary conditions, such as shape, mass and impinging velocity of projectile. Like any other simulation methods, FEM based ballistic test simulation too require some experimental validation to check the reliability of the final test results of simulation by cross checking it with real-time tests.

The current sub section of the chapter shall portray some of the essential key strategies, steps and notes taken by several researchers while solving such models.

4.1 Representative Volume Element (RVE)

A RVE or a unit cell is a micromechanical model of the composite structure which is used to predict the material behaviour, property and failure. An example of a RVE is shown in Fig. 4. It is the smallest unit of either a lamina and a portion of the composite which comprises of several elemental parameters such as fiber diameter, fiber layup and pitch, reinforcement particles, fiber volume fraction, possible void

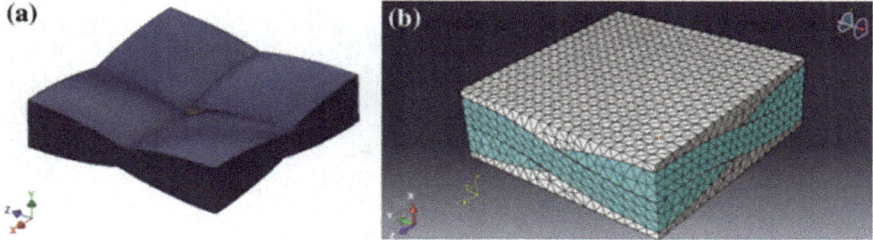

Fig. 4 Models of a unit cell **a** geometrical **b** meshed model (Pyka et al. 2017)

fracture and location, etc. By proper adjustments of these parameters, it is possible to get a very accurate estimate of the composite properties (Pyka et al. 2017).

The second important function of a RVE is the ease in modelling of complete composite structure. Unlike conventional materials, almost all composites are orthotropic in nature. This makes it very difficult to model them using any direct technique, even if one knows its orthotropic nature of properties. The RVE of the composite can be used in this process. Certain material engineering software such as Digimat and Inventor can make a lamina model from the RVE. This helps in creating a very accurate virtual model of the composite.

4.2 Geometry and Modelling

Non lamellar composite models are made directly from RVE filling up the entire dimensional volume of the composite body. In case of lamellar system, individual layers are stacked on top of each other keeping in mind of the fiber direction. Epoxy matrix is modelled with Cohesive Interaction property (CZM model) between each ply. As per this property, all the nodes, after meshing, of various layers in contact shall demonstrate interactive behaviour according to Traction-Seperation Law of desired order and Power Law, both of which shall determine the mechanisms of interlaminar damage initiation and propagation (Bodepati et al. 2017).

Although a lot of information and a visual data regarding the failure mechanism is received by modelling the full system, it often leads to high calculation time and requires more powerful solvers (Zarei et al. 2017). Due to this reason, under orthotropic and symmetric conditions, only 1/4th model of composite (Fig. 5) and projectile is made along the axes of symmetry (Yen 2002).

The shape of the projectile is given necessary attention. Sometimes, a very detailed model of projectile—a bullet is made which has a core and a stealth or casing. These are generally made other different metals or alloys as per the test condition. This projectile is placed is placed very close to the surface of the composite body ranging from a millimetre to a few micrometres for simulation purpose which is called gap interaction technique (Ansari and Chakrabarti 2017b). Sometimes a clay body is

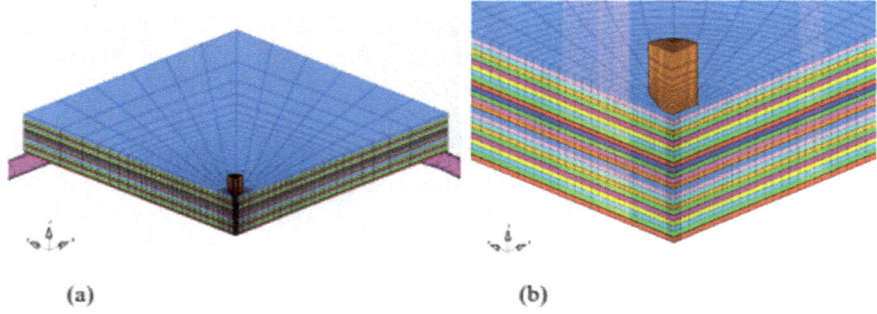

Fig. 5 Quarter of a composite modelled for simulation (Yen 2002)

modelled at the rear face of the panel which is made up of an isotropic material as per NIJ standards (Yang and Chen 2017).

4.3 Meshing and Boundary Conditions

Meshing and element choices are often very crucial to get reliable results. A lot of composite ballistic based research has been conducted using 3D 8 node hexahedral or quadrilateral (only one of these being used per entire model) for both composite body as well as the core of the projectile. As a rule of thumb, the fineness of the composite panel mesh should be higher than the projectile (Fig. 6). It should also be noted that reduction of calculation time can be obtained by increasing the fineness of primary yarn compared to secondary yarn as seen in Fig. 7. Primary yarn are those fibers which fall in line of impact requiring more focus while secondary yarns are the ones connected directly or indirectly to the primary yarns. High mesh fineness

Fig. 6 Comparison between mesh fineness of projectile and composite panel (Cao et al. 2015)

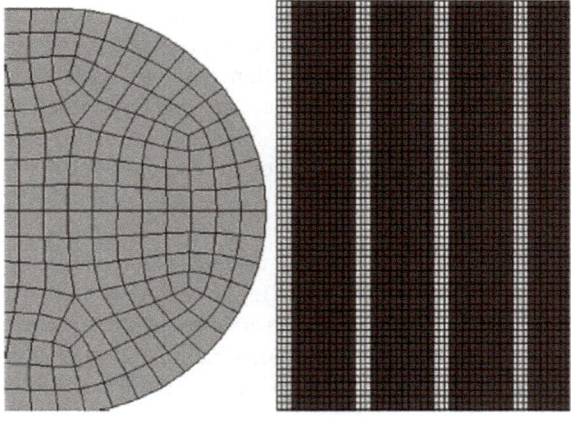

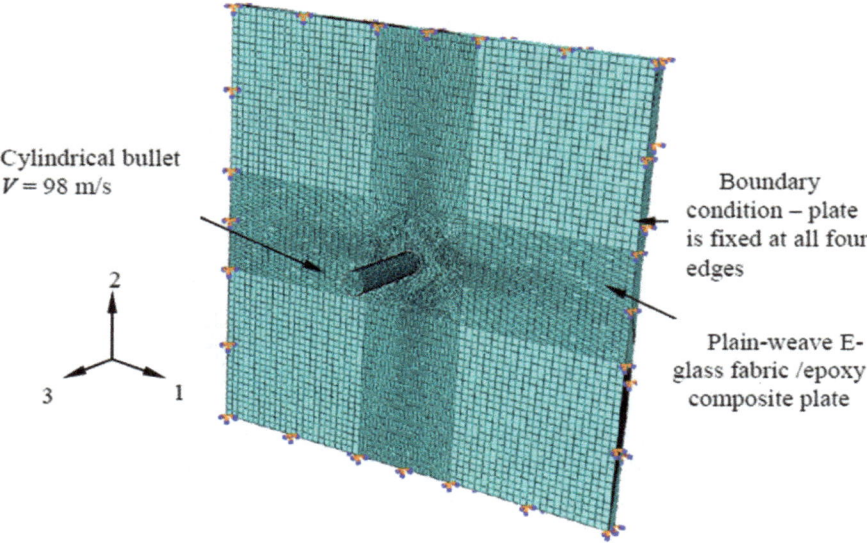

Fig. 7 Mesh fineness for primary and secondary fibers (Phadnis et al. 2013)

is implemented in the volume close to the impact point of the projectile as shown in Fig. 7.

Based on experimental results and assumptions, either a rigid or a deformable projectile is chosen for the simulation. If the complete composite panel and projectile are modelled, then free boundary condition is applied at all the edges of the panel, otherwise symmetric boundary conditions are applied along all lines of symmetry. A higher scaling factor is utilized for sliding interface penalty so that there is minimal permeation of non-contacted nodes although it leads to higher convergence time.

4.4 Reduced Integration and Hourglass Effect

During meshing, a multi-node element can have any number of integration points which help in determining the stresses and strain in that element. This is obtained by integrating the values at these points. Generally, most accurate values are obtained at these integration point and begins to show deviation as we move away from these points in the element space. One might think about increasing the integration points in an element but it would lead to higher computational time. Therefore modified elements are used to decrease computational time while making sure convergence condition is met and results are fairly within the desired range of error. These

Fig. 8 Large and small extent of hour glassing based on mesh size (Castillo et al. 2015)

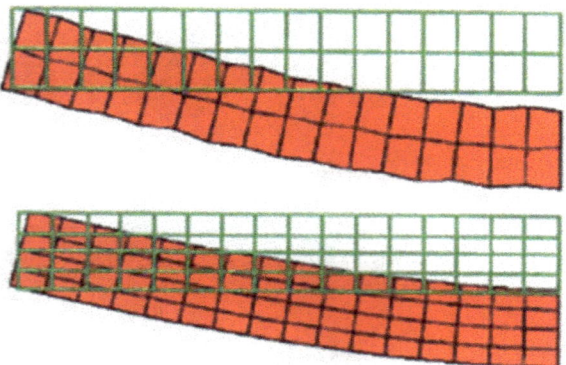

are called elements with reduced integration as they have less number of integration points such as S4R which is 4 node rectangular element (S4) having reduced integration—having only one integration point.

After post processing, there happens to be false deformation modes of the element as the material model appear to be distorted. This phenomenon is due to hourglass concept.

Hourglass effect are a set of zero-energy modes and non-physical deformation having no strain or stress. These modes occur only in under-integrated solid, shell, and thick shell elements. It tends to modify the structure's true response. If left alone, this may lead model rendering unstable and termination of the simulation (LS-DYNA Support 2019).

Although most of the results which had an hourglass effect can be visually identified, certain instabilities cannot be easily seen. Therefore, hour glassing energy is quantified with respect to the system energy. In ballistic testing of composites, the hourglass energy is commonly limited to 5% of the total energy for better accuracy. As opposed to loading individual nodes, pressure loading reduces hourglass excitation. Mesh refining also decreases this effect as seen in Fig. 8. Coarser mesh seems to have a greater hourglass effect as compared to finer mesh.

4.5 Application of Numerical Models

A FEM software is generally incorporated with various models which are described earlier in the chapter. After giving necessary boundary conditions to the model, a suitable numerical model out of the ones discussed is chosen for simulation. These numerical models are integrated with different solvers and subroutines as Finite Element (FE) codes. Depending upon the choice and necessity of the user, the appropriate solver and failure criterion is chosen. After meshing of the geometric model, the equations from numerical models are then applied at every nodal calculation along with an erosion algorithm. Based on the comparison between instantaneous geometric strain to a known level of equivalent geometric strain, an erosion algorithm removes highly distorted elements from calculation to reduce any irregularity in the results (Bandaru et al. 2016). The lost mass and momentum is re-distributed over other remaining nodes.

4.6 Result Interpretation and Post-processing

After the simulation is completed, the results can be viewed either by a visual deformation model or in a graphical representation. The stress and the deformation of the various layers of composite is obtained which can be compared with any other configuration of the composite or projectile. Other parameters such as DTL and ballistic limit of the simulated composite model can also be obtained. The ballistic limit (V_{bl}) is the minimum velocity with which a projectile can consistently penetrate the body completely of a given thickness (Berk et al. 2017). It can be obtained from the equation (Berk et al. 2017),

$$\frac{1}{2}(M+m)V_r^2 = \frac{1}{2}MV_i^2 - U_p \qquad (45)$$

where M and m are the projectile mass and the mass of material removed respectively, V_r is the residual velocity and V_i is the impact velocity. U_p is the penetration energy by specimen which is given by (Berk et al. 2017):

$$U_p = \frac{1}{2}MV_{bl}^2 \qquad (46)$$

The simulation result data can be fed to an image analysing software which can examine the number and the distribution (Cao et al. 2015). The crack propagation directions indicates the weak spots in the composite and predict the integrity of the panel.

The large amount of data can be fed to an Artificial Neural Network (ANN) which can help researchers to predict certain behaviour of the system such as ballistic limit of the composite as worked upon by (Artero-Guerrero et al. 2017).

5 Conclusion

The chapter discusses the numerical approach of evaluating ballistic characteristics of composite materials. A detailed depiction of the shortcomings of experimental methods which is overcome using analytical approach is given. Energy model, which is the most fundamental and basic approach for numerical analysis of an interactive system, is given a very comprehensive description. Every individual component of the general energy equation is explained with the mechanisms and the resulting equations. The key features and extensive narrative for each of these models and criterions are given along with some examples of work done by researchers using the same in their study. The application of FEM in ballistic tests on composites is mentioned along with the significance and challenges faced while working on the same. Every step or stage of the finite element analysis is described from the formulation of RVE and geometric modelling to result interpretation and post processing. Thus, this chapter gives an aspiring researcher an insight into the relevance of numerical approaches in analysing the performance of composite materials when subjected to ballistic impact.

References

Abrate S, Ferrero JF, Navarro P (2015) Cohesive zone models and impact damage predictions for composite structures. Meccanica 50(10):2587–2620. https://doi.org/10.1007/s11012-015-0221-1

Altair Engineering I (2020) Composites. Retrieved 19 April 2020 from https://insider.altairhyperw orks.com/composites-rupture/

Ansari MM, Chakrabarti A, Iqbal MA (2017) An experimental and finite element investigation of the ballistic performance of laminated GFRP composite target. Compos B Eng 125:211–226. https://doi.org/10.1016/j.compositesb.2017.05.079

Ansari MM, Chakrabarti A (2017a) Ballistic performance of unidirectional glass fiber laminated composite plate under normal and oblique impact. Procedia Eng 173:161–168. https://doi.org/10.1016/j.proeng.2016.12.053

Ansari MM, Chakrabarti A (2017b) Influence of projectile nose shape and incidence angle on the ballistic perforation of laminated glass fiber composite plate. Compos Sci Technol 142:107–116. https://doi.org/10.1016/j.compscitech.2016.12.033

Artero-Guerrero JA, Pernas-Sánchez J, Martín-Montal J, Varas D, López-Puente J (2017) The influence of laminate stacking sequence on ballistic limit using a combined experimental/FEM/artificial neural networks (ANN) methodology. Compos Struct 183(1):299–308. https://doi.org/10.1016/j.compstruct.2017.03.068

Balaganesan G, Velmurugan R, Srinivasan M, Gupta NK, Kanny K (2014) Energy absorption and ballistic limit of nanocomposite laminates subjected to impact loading. Int J Impact Eng 74:57–66. https://doi.org/10.1016/j.ijimpeng.2014.02.017

Bandaru AK, Chavan VV, Ahmad S, Alagirusamy R, Bhatnagar N (2016) Ballistic impact response of Kevlar® reinforced thermoplastic composite armors. Int J Impact Eng 89:1–13. https://doi.org/10.1016/j.ijimpeng.2015.10.014

Berk B, Karakuzu R, Toksoy AK (2017) An experimental and numerical investigation on ballistic performance of advanced composites. J Compos Mater 51(25):3467–3480. https://doi.org/10.1177/0021998317691810

Biegler MW, Mehrabadi MM (1995) An energy-based constitutive model for anisotropic solids subject to damage. Mech Mater 19(2–3):151–164. https://doi.org/10.1016/0167-6636(94)00015-9

Bodepati V, Mogulanna K, Rao GS, Vemuri M (2017) Numerical simulation and experimental validation of e-glass/epoxy composite material under ballistic impact of 9 mm soft projectile. Procedia Engineering 173:740–746. https://doi.org/10.1016/j.proeng.2016.12.068

Cao Y, Zhu S, Guo C, Vecchio KS, Jiang F (2015) Numerical investigation of the ballistic performance of metal-intermetallic laminate composites. Appl Compos Mater 22(4):437–456. https://doi.org/10.1007/s10443-014-9416-1

Castillo P, Pürgstaller A, Apostolidi E, Bergmeister K, Strauss A (2015) Innovative fiber reinforced elastomeric isolation devices. In: IABSE Conference, Structural engineering: providing solutions to global challenges—report. Geneva, pp 1637–1644. https://doi.org/10.2749/222137815 818359041

Chang FK, Chang KY (1987) Post-failure analysis of bolted composite joints in tension or shear-out mode failure. J Compos Mater 21(9):809–833. https://doi.org/10.1177/2F002199838702100903

Hashin Z (1980) Failure Criteria for Unidirectional Fiber Composites. J Appl Mech 47:329–334. https://doi.org/10.1115/1.3153664

Holmen JK, Hopperstad OS, Børvik T (2017) Influence of yield-surface shape in simulation of ballistic impact. Int J Impact Eng 108:136–146. https://doi.org/10.1016/j.ijimpeng.2017.03.023

Ismail MF, Sultan MTH, Hamdan A, Shah AUM, Jawaid M (2019) Low velocity impact behaviour and post-impact characteristics of kenaf/glass hybrid composites with various weight ratios. J Mater Res Technol 8(3):2662–2673. https://doi.org/10.1016/j.jmrt.2019.04.005

Karahan M, Jabbar A, Karahan N (2015) Ballistic impact behavior of the aramid and ultra-high molecular weight polyethylene composites. J Reinf Plast Compos 34(1):37–48. https://doi.org/ 10.1177/0731684414562223

Langston T (2017) An analytical model for the ballistic performance of ultra-high molecular weight polyethylene composites. Compos Struct 179:245–257. https://doi.org/10.1016/j.compstruct.2017.07.074

LS-DYNA Support. (2019). Retrieved 19 April 2019 from https://www.dynasupport.com/howtos/element/hourglass

Mallick PK (2007) Fiber-reinforced composites: materials, manufacturing, and design. CRC, Boca Raton

Mostafa NH, Ismarrubie ZN, Sapuan SM, Sultan MTH (2016) The influence of equi-biaxially fabric prestressing on the flexural performance of woven e-glass/polyester-reinforced composites. J Compos Mater 50:3385–3393. https://doi.org/10.1177/0021998315620478

Nayak N, Banerjee A, Panda TR (2017) Numerical study on the ballistic impact response of aramid fabric- epoxy laminated composites by armor piercing projectile. Procedia Eng 173:230–237. https://doi.org/10.1016/j.proeng.2016.12.002

Pach J, Pyka D, Jamroziak K, Mayer P (2017) The experimental and numerical analysis of the ballistic resistance of polymer composites. Compos B Eng 113:24–30. https://doi.org/10.1016/j.compositesb.2017.01.006

Pasquali M, Gaudenzi P (2017) Effects of curvature on high-velocity impact resistance of thin woven fabric composite targets. Compos Struct 160:349–365. https://doi.org/10.1016/j.compstruct.2016.10.069

Phadnis VA, Pandya KS, Naik NK, Roy A, Silberschmidt VV (2013) Ballistic impact behaviour of woven fabric composite: finite element analysis and experiments. J Phys: Conf Ser 451(1). https://doi.org/10.1088/1742-6596/451/1/012019

PNAE G-7-002-86 (1989) Equipment and pipelines strength analysis norms for nuclear power plants. Moscow, p 525

Pyka D, Jamroziak K, Blazejewski W, Bocian M (2017) Calculations with the finite element method during the design ballistic armour. In: Rusiński E, Pietrusiak D (eds) Proceedings of the 13th international scientific conference: computer aided engineering, pp 451–459. https://doi.org/10.1007/978-3-319-50938-9

Salman SD, Leman Z, Sultan MTH et al (2015) Kenaf/synthetic and kevlar®/cellulosic fiber-reinforced hybrid composites: a review. BioResources 10:8580–8603

Sobolev AV, Radchenko MV (2016) Use of Johnson-Cook plasticity model for numerical simulations of the SNF shipping cask drop tests. Nucl Energy Technol 2(4):272–276. https://doi.org/ 10.1016/j.nucet.2016.11.014

Tsai SW, Wu EM (1971) A general theory of strength for anisotropic materials. J Compos Mater 5:58–80

Yang Y, Chen X (2017) Investigation on energy absorption efficiency of each layer in ballistic armour panel for applications in hybrid design. Compos Struct 164:1–9. https://doi.org/10.1016/j.compstruct.2016.12.057

Yen CF (2002) Ballistic impact modeling of composite materials. In: Proceedings of the 7th international LS-DYNA users conference 6, pp 15–23. Retrieved from https://www.dynalook.com/international-conf-2002/Session_6-3.pdf

Zakikhani P, Zahari R, Sultan MTH, Majid DL (2016) Thermal degradation of four bamboo species. BioResources 11:414–425. https://doi.org/10.15376/biores.11.1.414-425

Zarei H, Sadighi M, Minak G (2017) Ballistic analysis of fiber metal laminates impacted by flat and conical impactors. Compos Struct 161:65–72. https://doi.org/10.1016/j.compstruct.2016.11.047

Zulkifli F, Stolk J, Heisserer U, Yong ATM, Li Z, Hu XM (2019) Strategic positioning of carbon fiber layers in an UHMwPE ballistic hybrid composite panel. Int J Impact Eng 129:119–127. https://doi.org/10.1016/j.ijimpeng.2019.02.005

Low Velocity Impact Testing and Post-impact Analysis Through Compression After Impact (CAI) and C-Scan

S. I. B. Syed Abdullah

Abstract The rapid increase in the use of composite materials within structural components has stimulated a significant amount of research in understanding the mechanics of the composite, as well as the ability to predict failure under a range of loading conditions. This includes impact, which can generate a considerable amount of damage in the composite. Low Velocity Impact (LVI) is one of the most critical load factors for composite laminates, incurring internal damage which could not be detected using the naked eye. This is particularly dangerous, since damage such as delamination may be present in the composite, which could seriously reduce the integrity of the structure. This chapter presents a review of the LVI testing, post-impact analysis and its subsequent response in Compression After Impact (CAI).

Keywords Low Velocity Impact (LVI) · Delamination · Compression After Impact (CAI) · Fracture toughness · Impact resistance · Impact damage · Impact performance · Impact toughness

1 Introduction

Low Velocity Impact (LVI) on composites is perhaps the commonest form of impact, presenting the most serious magnitude of damage, particularly for classical brittle fibre-brittle matrix composite systems, such as Carbon/Epoxy laminates. A considerable amount of delamination may appear in the laminate, significantly reducing the structure's integrity and load bearing capabilities.

Since the past few decades, a considerable amount of experimental testing has been focused on LVI. Large-mass, LVI often resulted in the so-called, Barely Visible Impact Damage (BVID), in which no clear visual of damage can be seen with the naked eye and requires additional detection tools such as the Non-Destructive Testing (NDT) techniques. In LVI, a bell-shaped load–deflection response almost always indicates that no damage is present in the laminate. In contrast, a sharp load drop will be observed at delamination onset, illustrated in Fig. 1b.

S. I. B. Syed Abdullah (✉)
Department of Aeronautics, Imperial College London, London, UK

© Springer Nature Singapore Pte Ltd. 2021
M. T. H. Sultan et al. (eds.), *Impact Studies of Composite Materials*, Composites Science and Technology, https://doi.org/10.1007/978-981-16-1323-4_12

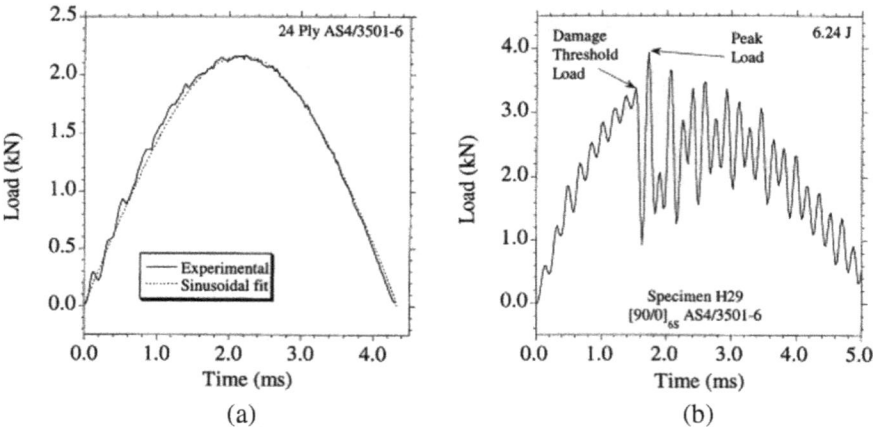

Fig. 1 Load-time response for24-ply GF/Epoxy laminate under two different LVI energies **a** 1.55 J, **b** 6.24 J (Schoeppner and Abrate 2000)

As mentioned earlier, when delamination is present, the load-bearing potential of the composite is significantly reduced, particularly its Compression after Impact (CAI) capabilities. The presence of delamination (usually impact-induced) will promote local buckling, eventually resulting in fibre rupture in the laminate. This effect may be more pronounced in a Quasi-Isotropic (QI) laminates, since delamination normally propagates in between plies with different orientation, with an increasing size for increasing misalignment angles (Davies and Olsson 2004).

Within the scope of damage characterisation, identification of damage types is usually done using Non-Destructive Techniques (NDT), such as the ultrasonic C-Scan and X-ray radiography methods. The latter are usually employed when identifying matrix and fibre failure, whilst the former on delamination damage. More advanced techniques such as fractographic analysis are usually performed if required, or to understand the progression of damage in a laminate.

In the following sub-sections, a discussion will be made with regards to damage characterisation of composites under LVI, with an emphasis on CAI and its associated damage characterisation techniques.

2 Low Velocity Impact on Composite Structures

In general, LVI is defined as an impact event in which through-thickness stress distribution play no significant part in the stress distribution at any time during the impact event. at any time during the impact event. A global deformation is usually observed in the laminate, Fig. 2 due to the long impact duration.

The contact stress between the projectile (or impactor) and the laminate would generally induce minimal damage in the form of matrix cracking (Cantwell and

Fig. 2 Boundary dominated
event (Davies and Olsson
2004)

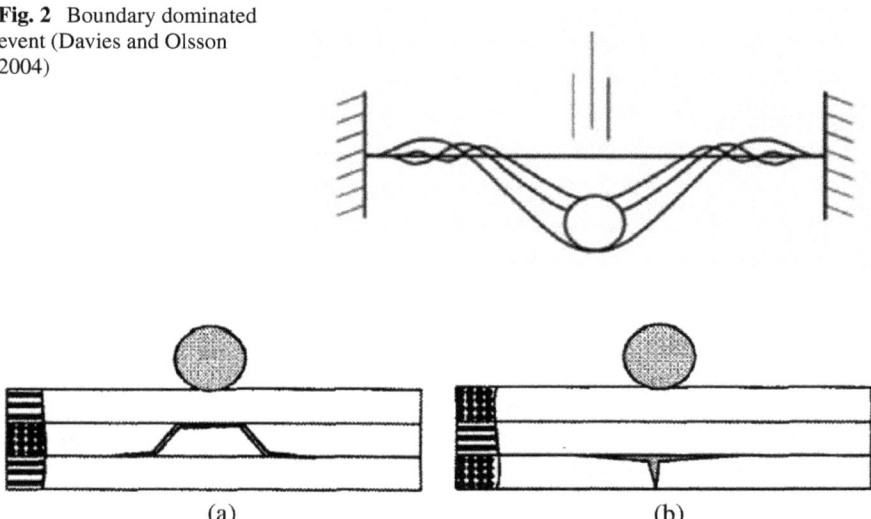

(a)   (b)

Fig. 3 Types of matrix failure **a** Tensile crack, **b** Shear crack (Abrate 1998)

Morton 1989). These cracks would saturate as a result of coalescence between
multiple microcracks (Olsson 2001; Berthelot 2003; Williams et al. 2003; Puck
and Schürmann 2004). Depending on the thickness of the laminate, the damage
pattern would differ due to the difference in the energy absorbing mechanism. For
a typical brittle fibre-matrix system such as Carbon Fibre/Epoxy (CF/EPoxy) and
Glass Fibre/Epoxy (GF/Epoxy), the cross-sectional damage normally resembles the
so-called "pine tree" pattern. For thin laminates, a reversed pine tree pattern, Fig. 4b,
can usually be observed from the cross-section of the impact area. This is due to the
high bending load at the rear side of the laminate, hence initiating shear matrix failure,
Fig. 3b. On the contrary, matrix cracks in thick laminates will often result in a pine
tree pattern, Fig. 4a. This is because the normal in-plane stresses have exceeded the
transverse tensile stress of the front plies, therefore initiating matrix failure. For both
pine tree patterns, extensive experimental evidence exists for CF/Epoxy composites
(Cantwell and Morton 1985; Jih and Sun 1993; De Freitas et al. 2000; Bouvet et al.
2009; García-Rodríguez et al. 2018), and GF/Epoxy (Zhou 1995; Shyr and Pan 2003;
Liaw and Delale 2007; Crupi et al. 2014).

2.1   Compression After Impact

As mentioned in the previous section, LVI normally induces the so-called BVID,
which is particularly insidious since it can have a large effect on the laminate residual
strength yet having little to no clear evidence of damage on the laminate surface.

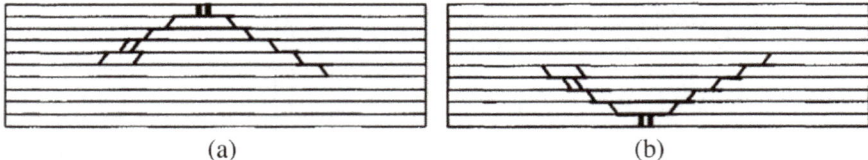

Fig. 4 **a** Pine tree pattern, **b** Reversed pine tree pattern (Abrate 1998)

Inside the laminate, extensive internal delamination may be found by using ultrasonic techniques such as the C-scan. An image of an internal delamination of an impacted QI GF/Epoxy specimen is shown in Fig. 5, in which extensive internal delamination can be observed forming a pine-tree pattern induced from LVI loads.

The presence of delamination has been reported to reduce the residual compressive strength up to 70% (Davies and Olsson 2004). This further exacerbates the composite's structural integrity by promoting local buckling of sub-laminates in the

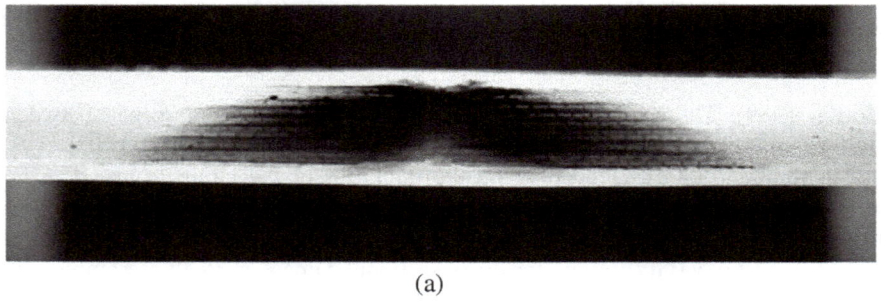

(a)

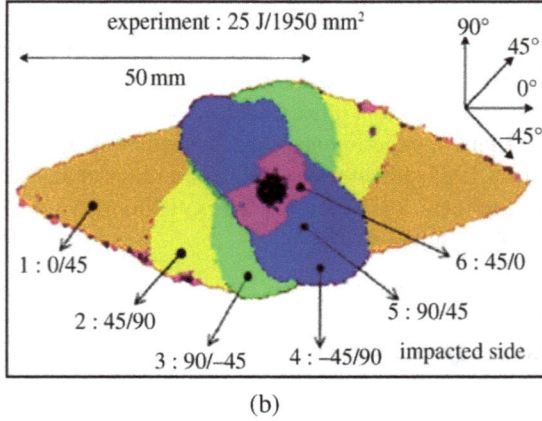

(b)

Fig. 5 Extent of damage on an impacted QI GF/Epoxy laminate **a** Edge view, **b** Top view generated from C-scan (Wisnom 2012)

early stages of CAI loading (Abdulhamid et al. 2016), ultimately resulting in a further propagation of delamination before laminate total collapse. Depending on the type of delamination, a different CAI response may be observed in the laminate. In addition, the stacking sequence, as well as the ply orientation also have a significant effect in determining the laminate residual strength from impact-induced loads (Amaro et al. 2014). Fabric architecture also plays an important role in determining the impact residual strength. For instance, a woven architecture is generally more delamination-resistant compared to UD-based architecture. This is because the crack growth is limited in woven-based architecture due to the fibre crimp, diverting the crack from propagating on its original path. Thus, more energy is required to propagate the crack (Greenhalgh 2009).

2.2 Low Velocity Impact and Compression After Impact Testing

In general, the LVI testing is performed using an instrumented drop tower, equipped with a photoelectric sensor to measure the impact velocity. The impactor, usually hemispherical in shape, is installed with a load cell at the tip to capture the impact force exerted on the laminate. The drop tower is also fitted with a pneumatically operated anti-rebound system to prevent multiple impacts on the target laminate. Figure 6a and b presents a typical instrumented drop tower and the impactor, respectively.

The Boeing/ASTM D7136 (ASTM-D7136-15 2015) and D7137 (ASTM D 7137 2012) is commonly used as a guideline in performing the LVI testing and the CAI testing, respectively. Composite laminates, usually 100×150 mm with varying thickness are impacted at the plate mid-point at specific energy levels to induce damage. Upon impact, the damaged laminates are then placed in a standard CAI fixture which is then loaded in compression until final laminate failure. Figure 7a and b presents an illustration of the typical laminate dimension, and the CAI test fixture, respectively.

3 Damage Characterisation

Non-destructive techniques, such as the ultrasonic C-scan and X-ray radiography are usually employed to characterise the types of damage induced from impact. The C-scan uses high-frequency sound waves beamed into the material to detect subsurface defects or discontinuities. The sound waves then travel through-the-thickness f the material, and then reflected back at the interfaces, such as changes in ply direction or planar defects. The reflected beams are then analysed using information such as the time of flight, to deduce the location (depth) of any flaws or discontinuities (Greenhalgh 2009).

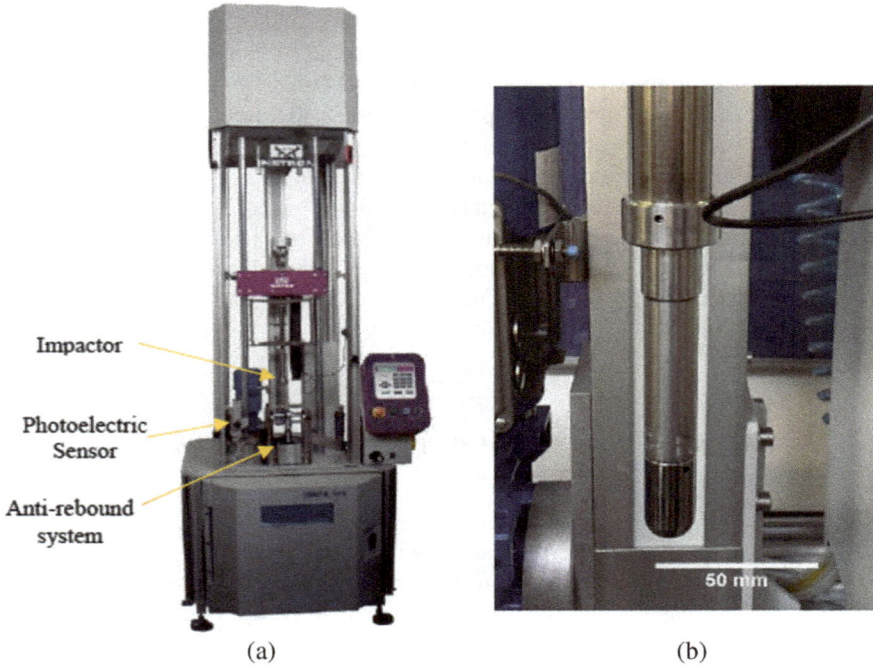

(a)   (b)

Fig. 6 Typical LVI testing setup **a** Drop tower, **b** Impactor

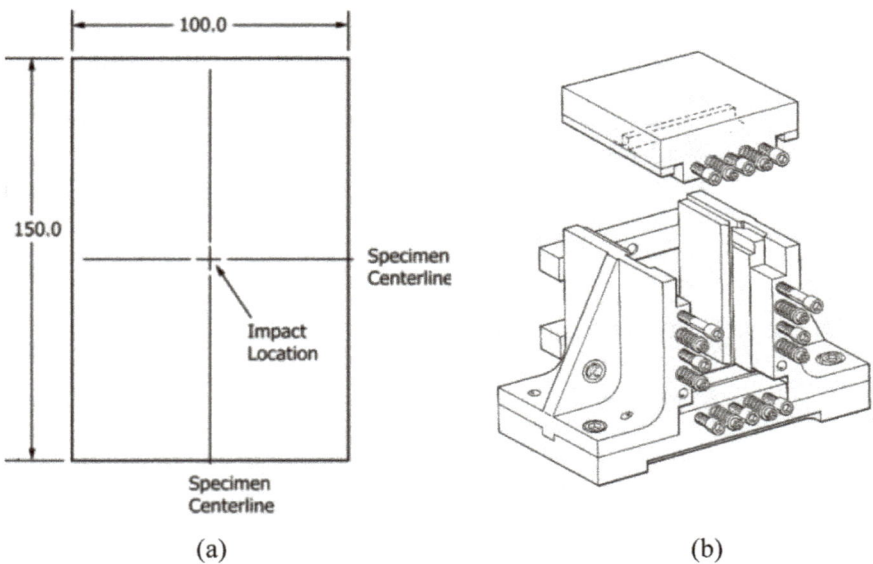

(a)   (b)

Fig. 7 **a** Typical laminate dimensions in LVI testing, **b** CAI test fixture

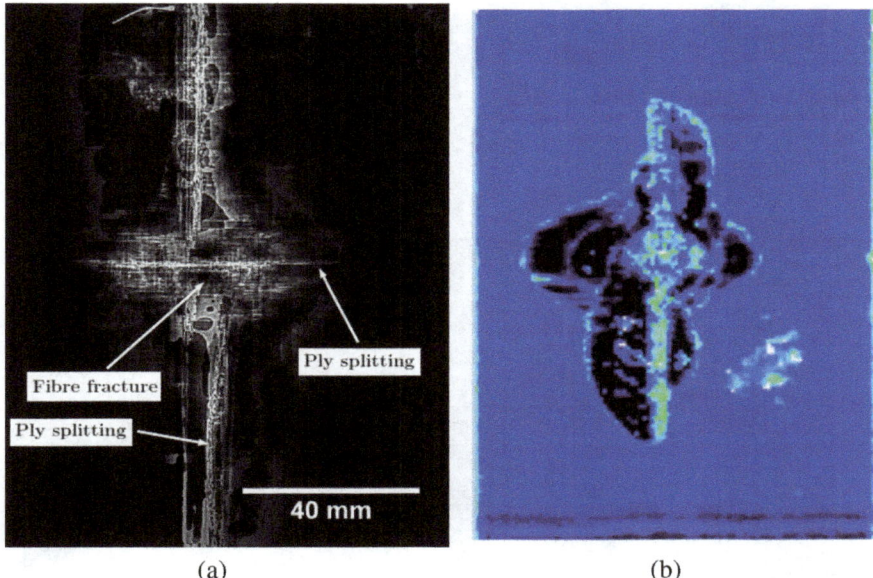

Fig. 8 Characterisation of damage on a CF/Epoxy laminate under 40 J of LVI load using **a** X-ray radiography, **b** Ultrasonic C-scan

The X-ray radiography technique usually employs the use of a dye-penetrant to enhance the image produced. In polymer composites, the low atomic weight of all the elements in the material (carbon, oxygen, nitrogen and hydrogen) means that defects will not attenuate radiation much more than the pristine composite, and thus, any defects are difficult to resolve (Greenhalgh 2009). It is often used to capture the damage associated with matrix cracks and fibre fracture. Figure 8a and b presents damage characterised using X-ray radiography and the ultrasonic C-scan technique. Note that the X-ray radiography captures mainly fibre fracture, and matrix failure, whilst the C-scan mainly captures delamination damage.

In CAI, damage is usually more extensive, with delamination extending towards the laminate boundaries. This is particularly true for thick laminates, whereas for thin laminates, a buckling deformation tended to induce the propagation of delamination transverse to the loading direction at lower load levels. Naturally, thinner laminates will have a lower CAI strength if compared to the thicker laminates. During CAI, damage in the form of fibre fracture (due to the local ply buckling propagates from the impact site towards the laminate boundaries. This is shown in Fig. 9, where it can be observed that extent of damage prior to CAI testing.

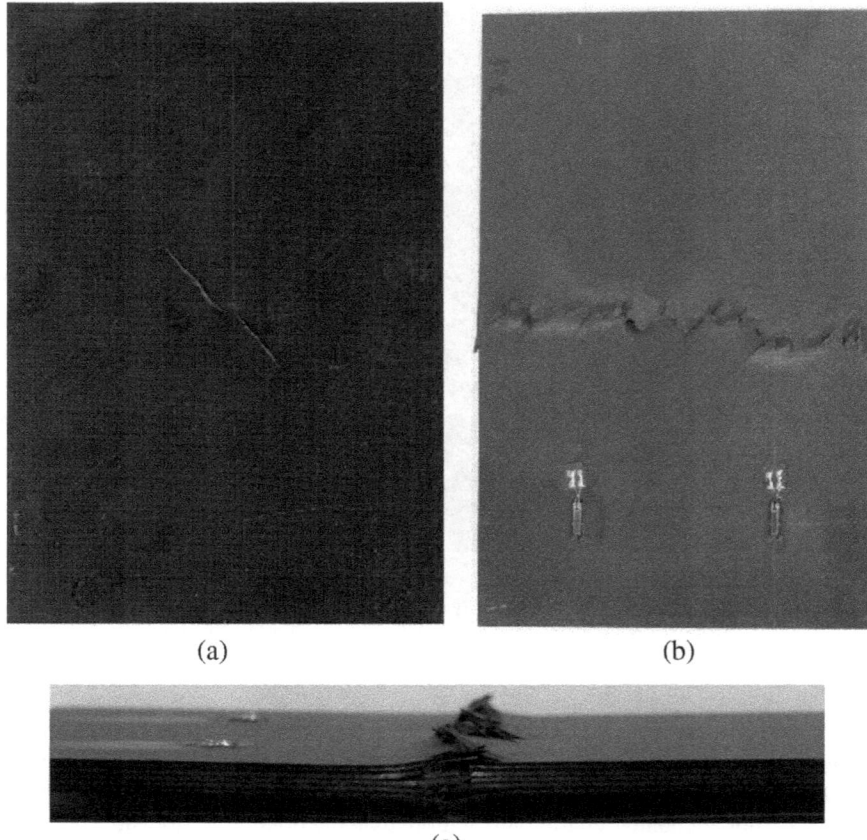

(a)   (b)

(c)

Fig. 9 Damage on the front surface of CF/Epoxy laminates **a** post-impact, **b** post-CAI testing, **c** side view showing fibre failures and ply buckling

4 Conclusion

Impact induced damage, particularly LVI damage, can present the most serious in nature, due to the hidden nature of damage inside the laminate, whilst minimal signs of damage can be on the laminate surface. It is possible that the laminate suffers extensive amount of internal delamination, which severely degrades its load bearing capabilities.

Detection of LVI damage is usually performed with the help of NDT techniques, such as the ultrasonic C-scan and the X-ray radiography. While there may be more advanced techniques, such as using Acoustic Emission (Saeedifar et al. 2020), the C-scan and X-ray techniques offers a relatively quick approach in characterising damage due to impact and CAI. The progression of damage can also be understood, though this may require a more 'expensive' techniques such as fractography and optical microscopy, since the laminate has to be dissected at certain section of interest, and that placed under a microscope to understand the damage behaviour.

In recent years, considerable efforts have been made to improve the LVI response and the CAI behaviour, hence ultimately improving the structural crashworthiness and load bearing capabilities. These include the use of different fabric architecture, polymer fibre based composite systems and thermoplastic matrices due to the inherent toughness in terms of delamination. When delamination is supressed, the CAI response may be improved since delamination promotes local ply-buckling.

References

Abdulhamid H, Bouvet C, Michel L et al (2016) Experimental study of compression after impact of asymmetrically tapered composite laminate. Compos Struct 149:292–303. https://doi.org/10.1016/j.compstruct.2016.04.015

Abrate S (1998) Impact on Composite Structures. Impact Compos Struct. https://doi.org/10.1017/cbo9780511574504

Amaro AM, Reis PNB, de Moura MFSF, Neto MA (2014) Buckling analysis of laminated composite plates submitted to compression after impact. Fibers Polym 15:560–565. https://doi.org/10.1007/s12221-014-0560-x

ASTM-D7136-15 (2015) Standard test method for measuring the damage resistance of a fiber-reinforced polymer matrix composite to a drop-weight impact event. ASTM Int

ASTM D 7137 (2012) Standard test method for compressive residual strength properties of damaged polymer matrix composite plates. ASTM Int, pp 1–17. https://doi.org/10.1520/D7137_D7137M-17

Berthelot JM (2003) Transverse cracking and delamination in cross-ply glass-fiber and carbon-fiber reinforced plastic laminates: static and fatigue loading. Appl Mech Rev 56:111–147. https://doi.org/10.1115/1.1519557

Bouvet C, Castanié B, Bizeul M, Barrau JJ (2009) Low velocity impact modelling in laminate composite panels with discrete interface elements. Int J Solids Struct 46:2809–2821. https://doi.org/10.1016/j.ijsolstr.2009.03.010

Cantwell WJ, Morton J (1989) Comparison of the low and high velocity impact response of CFRP. Composites 20:545–551. https://doi.org/10.1016/0010-4361(89)90913-0

Cantwell WJ, Morton J (1985) Detection of impact damage in CFRP laminates. Compos Struct 3:241–257. https://doi.org/10.1016/0263-8223(85)90056-X

Crupi V, Epasto G, Guglielmino E (2014) Internal damage investigation of composites subjected to low-velocity impact. Exp Tech. https://doi.org/10.1111/ext.12096

Davies GAO, Olsson R (2004) Impact on Composite Structures. Aeronaut J 108:541–563. https://doi.org/10.1017/S0001924000000385

De Freitas M, Silva A, Reis L (2000) Numerical evaluation of failure mechanisms on composite specimens subjected to impact loading. Compos B Eng 31:199–207. https://doi.org/10.1016/S1359-8368(00)00003-2

García-Rodríguez SM, Costa J, Bardera A et al (2018) A 3D tomographic investigation to elucidate the low-velocity impact resistance, tolerance and damage sequence of thin non-crimp fabric laminates: effect of ply-thickness. Compos Part A Appl Sci Manuf 113:53–65. https://doi.org/10.1016/j.compositesa.2018.07.013

Greenhalgh ES (2009) Failure analysis and fractography of polymer composites

Jih CJ, Sun CT (1993) Prediction of Delamination in Composite Laminates Subjected to Low Velocity Impact. J Compos Mater 27:684–701. https://doi.org/10.1177/002199839302700703

Liaw B, Delale F (2007) Hybrid carbon-glass fiber/toughened epoxy thick composite joints subject to drop-weight and ballistic impacts

Olsson R (2001) Analytical prediction of large mass impact damage in composite laminates. Compos Part A Appl Sci Manuf 32:1207–1215. https://doi.org/10.1016/S1359-835X(01)00073-2

Puck A, Schürmann H (2004) Failure analysis of FRP laminates by means of physically based phenomenological models. Fail Criteria Fibre-Reinf-Polym Compos 832–876. https://doi.org/10.1016/B978-008044475-8/50028-7

Saeedifar M, Saleh MN, El-Dessouky HM, et al (2020) Damage assessment of NCF, 2D and 3D woven composites under compression after multiple-impact using acoustic emission. Compos Part A Appl Sci Manuf 132. https://doi.org/10.1016/j.compositesa.2020.105833

Schoeppner GA, Abrate S (2000) Delamination threshold loads for low velocity impact on composite laminates. Compos Part A Appl Sci Manuf 31:903–915. https://doi.org/10.1016/S1359-835X(00)00061-0

Shyr TW, Pan YH (2003) Impact resistance and damage characteristics of composite laminates. Compos Struct 62:193–203. https://doi.org/10.1016/S0263-8223(03)00114-4

Williams KV, Vaziri R, Poursartip A (2003) A physically based continuum damage mechanics model for thin laminated composite structures. Int J Solids Struct 40:2267–2300. https://doi.org/10.1016/S0020-7683(03)00016-7

Wisnom MR (2012) The role of delamination in failure of fibre-reinforced composites. Philos Trans R Soc A Math Phys Eng Sci 370:1850–1870. https://doi.org/10.1098/rsta.2011.0441

Zhou G (1995) Prediction of impact damage thresholds of glass fibre reinforced laminates. Compos Struct 31:185–193. https://doi.org/10.1016/0263-8223(94)00062-X

Low Velocity Impact Characterization of Flax/Kenaf/Glass Fibre Reinforced Epoxy Hybrid Composites

Noorshazlin Razali, M. T. H. Sultan, A. U. M. Shah, and S. N. A. Safri

Abstract This study characterized low velocity impact behaviour of flax/kenaf/glass woven fibre reinforced epoxy hybrid composites. Hybrid composites were fabricated by hand lay-up technique while maintaining a total fibre loading 40 wt%, and ratio of hybridization of flax woven fibre, kenaf woven fibre and glass woven fibre was 15:35:50. Four types of glass woven fibre were used which were E200, E400, E600 and E800. The low velocity impact tests were conducted by using IM10 Drop Weight Impact Tester. The impact energy was varied at 15, 30, 45, 60, 75 and 90 J for each type of specimens. Three repeatability tests were carried out for each type of specimens and for each impact energy level. The specimens later on undergo the post impact test in order to analyse the failure obtained from the impact. The results show that the E600 specimens had the highest energy absorption. The results from this study might be used for future research using flax, kenaf and glass woven fibre composites.

Keywords Low velocity · Bio-composite · Natural fiber · Flax · Kenaf · Glass · Hybrid composites · Characterization

1 Introduction

The advancement of the up and coming age of materials that can be considered as elective items and procedures are increasing more consideration. This is the consequence of the joining of worldwide environmental factors together with the standards of manageability and modern nature that show eco-effectiveness and green building.

N. Razali · M. T. H. Sultan (✉) · A. U. M. Shah
Department of Aerospace Engineering, Faculty of Engineering, Universiti Putra Malaysia, 43400 UPM Serdang, Selangor, Malaysia
e-mail: thariq@upm.edu.my

M. T. H. Sultan · A. U. M. Shah · S. N. A. Safri
Laboratory of Biocomposite Technology, Institute of Tropical Forestry and Forest Products, Universiti Putra Malaysia, 43400 UPM Serdang, Selangor, Malaysia

M. T. H. Sultan
Aerospace Malaysia Innovation Centre (944751-A), Prime Minister's Department, MIGHT Partnership Hub, Jalan Impact, 63000 Cyberjaya, Selangor Darul Ehsan, Malaysia

© Springer Nature Singapore Pte Ltd. 2021
M. T. H. Sultan et al. (eds.), *Impact Studies of Composite Materials*, Composites Science and Technology, https://doi.org/10.1007/978-981-16-1323-4_13

Composite materials, particularly green composites, ideally and unquestionably fit well into this new change in perspective (Mohanty et al. 2005). All the fibres utilized in composite materials can be separated into two classifications, to be specific natural fibres and synthetic fibres. In the current days, composites are created from a little scope to a bigger scope upon its process with levels of popularity and extraordinary supplies. Research interests in natural fibres and synthetic fibres are developing in the composite material field. Synthetic fibres are the most commonly used fibres nowadays. Further research and development on natural fibres should be enhanced, so that the dependency on synthetic fibres could be reduced due to its limited sources and environmental endangerment.

Impact can be described as a crash between at least two things, which can be flexible, liquid, plastic, or any blend of these for its impact reaction (Stronge 2018). The impact reaction is crucial to the investigation of impact elements of crack and fracture (Borvik 2003). Impact likewise starts a delamination in the composites (Razali et al. 2018). A delaminated composite will have a diminished residual compressive quality that can prompt to disastrous disappointment to any structures. Other than that, delamination is hard to distinguish by visual assessment and, accordingly, requires further developed systems for its identification (Kaw 2006). In composites, the absorption of the impact can cause an enormous harmed region and reduced the quality and solidness of those composites. The reaction to impact on composites can be brought about by various properties of the composites, for example, the sort of fiber, thickness of the composites, the hybridization of the fiber, and furthermore the lay-up arrangement and direction (Tirillo et al. 2017).

Impact occasion of low-velocity impact is happening in the scope of 1–10 m/s. The impact occasion is reliant on the material properties, target firmness, and the penetrator solidness and mass (Balali et al. 2017). The impact damage brought about by low-velocity impacts is extremely perilous in light of the fact that the damage could be unidentified by the unaided eye and can prompt an unexpected failure of the composite materials. A low-velocity impact test can be studied by utilizing a drop-weight test or a Charpy test. A drop-weight impact testing is one of the most widely recognized tests for low velocity impact damage in composite materials (Duell 2014). In a drop-weight test, an overwhelming mass is utilized to create a kinetic energy those outcomes in an effect on the test samples at the low-velocity impact region. The height of the striker is controlled relying upon the force and impact energy applied on the test samples. Low-velocity impact testing is conceivably perilous in light of the fact that the damage caused may be left unidentified (Channabasavaraju et al. 2013). The damage is viewed as shaped of low-energy conditions when the impactor's speed is under 40 m/s (Ricci et al. 2013).

One of the other kinds of low-velocity impact tests is by utilizing a swinging pendulum. A swinging pendulum impact test also known as a Charpy Test, comprises of a turning inflexible arm with a projectile head, or a projectile mass, swinging on links. The projectile swings into a test sample that is normally mounted in the vertical plane. The projectile is discharged from a foreordained stature comparing to an ideal impact energy level. The falling weight is typically guided by a cylinder or rail framework in order to accomplish better focusing on the target samples (Agrawal et al. 2014).

A dropped device in low-velocity impacts test can be depicted by the main method of vibration, or the static mode characterization. Higher modes can be disregarded in light of the fact that the contact force term is any longer than the time required for the impact wave to arrive at the limits and back to initial. This can be approximated by an energy balance model where the absolute energy of the framework is monitored, dismissing the energy of much more vibration modes, erosion, and different misfortunes (Kim et al. 2012).

Nowadays, an extraordinarily structured vertical drop-weight test was utilized to supplant the current drop test. The projectile utilized was a 12 mm distance across hemispherical tup nose projectile. The samples of the test were held with its holds, and the cylinders were fit for applying in-plane stacking in two autonomous opposite axes on the samples. The aftereffect of this experiment indicated that clipping the sample at its four sides brought about an increasingly steady structure, contrasted with a two-side clipping (Akin and Senel 2010). Rafiq et al. (2017) directed a drop-weight impact test on a glass fibre and nanoclay material reinforced epoxy hybrid composites with the impact energies somewhere in the range of 10 and 50 J. In their studied, they examined the impact of nanoclay expansion on glass fibre hybrid composites reinforced with epoxy, and whether it can improve the properties of the glass fibre. It was determined that the hybrid composites had a lower impact damage contrasted with the glass fibre composites itself. Rahman et al. (2015) likewise led an analysis on low-velocity impact and contemplated the impact properties of carbon nanofibres incorporated with carbon fibre reinforced epoxy produced by the Out of Autoclave and Vacuum Bag Only (OOA-VBO) method. The impact was examined by utilizing scanning electron microscopy (SEM) and the ultrasonic C-scan. It was seen that acceptable grip between the carbon nanofibres and the polymers has improved the interfacial holding between the matrix and fibre, prompting a higher impact opposition of the composites.

The varieties in impact behaviour in a low-velocity impact testing are because of the distinction in the measurements of the sample. Test design and stacking sequence didn't influence the size of the delamination territory under fluctuating energy absorb and the highest contact force (Ghelli and Minak 2011). The samples was explored with various thicknesses of layers being overlaid together to examine the impact behaviour.

In this study, low velocity impact test has been chosen to conduct a test on hybrid of Kenaf fibre, flax fibre and glass fibre since this fibre were used in various applications. Previous researchers have conducted studies on the comparison between different types of composite. Among the natural fibres being explored, kenaf and flax fibres are currently most commonly used in industries in well-established applications (Pickering et al. 2016). Research involving kenaf and flax fibres has also witnessed increased interest, mainly due to their attractive properties and abundant availability in Malaysia (Baley et al. 2018; Tholibon et al. 2019). Kenaf composites itself is well known has lower properties as compared to other natural fibre. Moreover, the studies of low velocity on natural and synthetic fibre such as kenaf, flax and glass fibre have not been done before. Therefore, this study needs to be done in order to analyse and to study the impact behaviour of those natural and synthetic hybrid composites.

Kenaf fibre is well known in Malaysia and easy to get here. To improve the properties of the kenaf, flax has been chosen to hybridise with kenaf as flax is well known that has good properties as biocomposites. Glass fibre has been chosen as the synthetic material that will hybridise with the flax/kenaf composites. Glass fibre is less expensive than carbon fibre and Kevlar. Due to the cost of production nowadays, this material is chosen to test its strength and its impact behaviour in low velocity impact. The main interest in this research is to compare the four different type of glass fibre in terms of impact resistance, stiffness, and toughness that impacted with different impact energy level. While doing the low velocity impact test, acoustic emission data were recorded in order to study the impact damage occurs in the composite. Finally, at the end of this research, conclusions can be summarize about the type of these materials which are safe to be implemented in inner structural applications as a replacement for existing materials (as examples type S-glass and Kevlar) because of their high cost, materials availability and not environmental friendly.

2 Experimental

2.1 Fabrication of Composite Specimen

Flax, kenaf and glass woven cloth fibre were chosen as fibre reinforcement for this research. There were four types of glass fibre that had been used which were E200, E400, E600 and E800. The cloth fibres materials and epoxy resin were purchased from ZKK Sdn. Bhd. The specimens were prepared in the laboratories of the Faculty of Engineering, Universiti Putra Malaysia (UPM, Seri Kembangan, Malaysia). Specimens were prepared based on to the following sequence: fibre preparation, lay-up process, curing process, and cutting to a standard size of low velocity impact samples using a CNC machine at Material Forming Laboratory, Universiti Putra Malaysia.

Flax, kenaf and glass fibre woven mats were cut into 350 × 350 mm sizes. The hybrid composites were fabricated with all woven fibres aligned in 90° angle of orientation. A square plate aluminium mould was used for the fabrication process. Acetone was used to clean the mould of any residues. Wax was used as a releasing agent to prevent specimens from sticking to the mould after the curing process. Epoxy resin brand Zeepoxy HL002TA with hardener brand Zeepoxy HL002TB were used as the matrix in current study. Flax and kenaf fibres were laid up in a 30:70 weight ratio while maintaining a 40:60 ratio between fibre and resin. While the ratio between flax/kenaf fibre with glass fibre was 50:50. An aluminium plate coated with wax was placed on top of the composite lay-up and loads were placed to ensure the uniform distribution of resin. The material was cured for 48 h, followed by post-curing in an oven at 80 °C for 2 h, before it was cut into standard size specimens.

2.2 Low Velocity Impact Test

To do the low-velocity impact testing, an IM10 Drop Weight Impact Tester was used. This test was done at the Laboratory of Aerospace Engineering, Faculty of Engineering, Universiti Putra Malaysia. The total of weight of the impactor was 5.101 kg and the striker weight is 0.325 kg. The samples were grip by the clamping unit below the drop weight impactor to prevent any movement during the test. The striker was impacted at the centre of the samples. The impact energy using in this research was 15, 30, 45, 60, 75 and 90 J for each type of specimens. Three repeatability tests were carried out for each type of specimens and for each impact energy level. The height of the drop weight will vary the impact energy value and also the velocity of the impactor. In order to get a one drop impact test, the machine was set up anti-rebound mass.

The data of displacement, impact force, impact energy and time was then acquired using the Imatek Impact Analysis software that connected to the IM10 Drop Weight Impact Tester. The weight carriages were ensured in beating the rest in the Safe Park position. Then, the correct striker and correct mass was fitted at the Mass Carriage. The total mass which was the dropped mass and striker mass was entered into the Data Capture Window, at the 'drop mass' field. The total striker mass which were the striker mass and half transducer was then entered into the Data Capture Window, at the 'striker mass' field. The specimen on the anvil was positioned and secured and an impact chamber access door was closed. The impactor weight was then raised a bit higher to clear the safety area. To set the position sensor, the safety door was opened so that the weight can be moved freely. The impactor weight was lowered onto the specimen and the Trigger Point (zero point) was set. The impactor anvil then being adjusted until to the desired drop height for certain impact energy level. The height required for certain impact energy was calculated using Eq. (1).

$$Impact\ energy,\ E_i = mgh \qquad (1)$$

m = impactor mass.
g = gravitational force.
h = impactor height.

Table 1 shows the height needed to varies before the drop weight impactor can be released for impact for a certain impact energy. All the data needed was set in the software. Finally, the 'Release' button was operated. The energy absorbed by the impacted specimens was determined by calculating the area under the curve of Force versus Displacement.

Table 1 Height of drop weight impactor release for certain impact energy	Impact energy (J)	Height of the drop weight impactor (m)
	15	0.3
	30	0.6
	45	0.9
	60	1.2
	75	1.5
	90	1.8

2.3 Compression After Impact

Following the ultrasonic C-scan, all of the specimens underwent the compression after impact (CAI) test to determine the residual compressive strength of the specimens. The impacted specimens were subjected to an in-plane compression testing. The test was done according to the anti-buckling Boeing CAI test based on the ASTM D 7137 standard. A 300 kN load cell universal testing machine was used to conduct the test. A displacement rate of 1.25 mm/min was set. The load and displacement of the compression after impact test were recorded. The ultimate compressive residual strength was determined using Eq. (2).

$$RS_{uc} = \frac{F_u}{wt} \tag{2}$$

RS_{uc} = ultimate compressive residual strength.
F_u = ultimate force.
w = width of the specimens.
t = thickness of the specimens.

3 Results and Discussion

3.1 Low-Velocity Impact

Six different impact energy levels were used in this test in order to study the penetration behaviour of the specimens and the difference between them. Three repeatability tests have been done for each type of the specimens and each impact energy levels. The data of the three average values were recorded as shown in Figs. 1, 2 and 3.

Figure 1 illustrates the force versus time graph of the low-velocity impact test for the hybrid composites (glass/kenaf/Flax BL150), named after the type of glass fibre used. The impact energy levels that were used to conduct the experiments ranged from 15 J up to 90 J. The reaction force from the specimen to the impactor was defined from force–time curves. This graph shows the damage progression from the initial

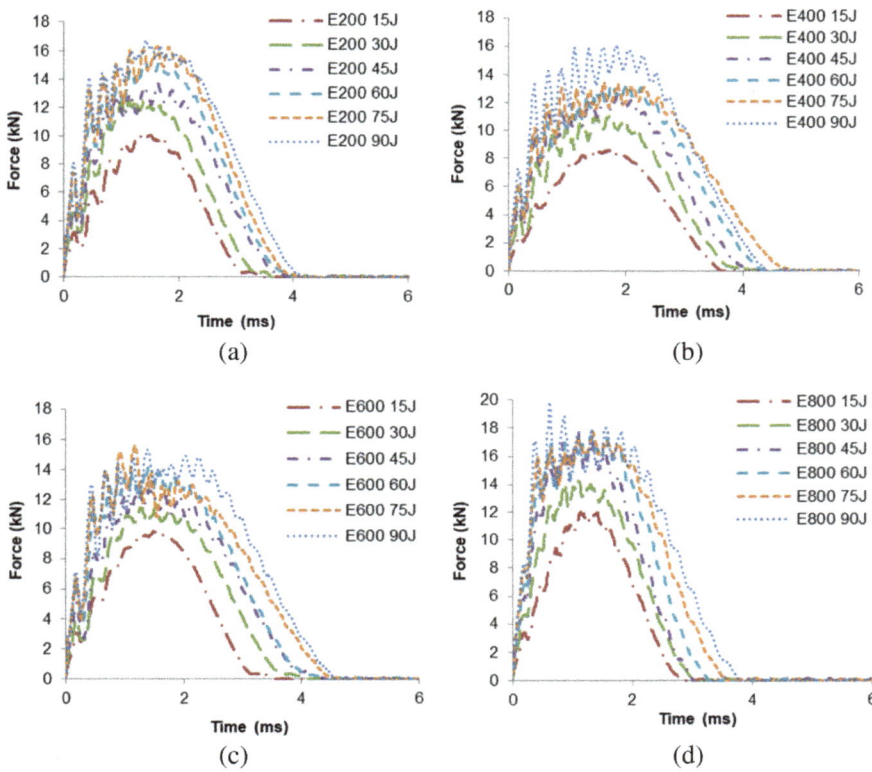

Fig. 1 Force–time curves of the flax-kenaf fibre composite hybridised with glass fibre: **a** E200; **b** E400; **c** E600; and **d** E800

force until the damage growth. The contact force during impact showed significant oscillations. Few researchers have suggested that these oscillations are usually due to the elastic wave responses and the vibration of the test samples. From the graph, we can see that as the peak force increased, the time increased as well. The shape of the curves shows that the fracture of the specimen had occurred. The graph increased until the highest force and decreased to a zero-force, indicating that a failure in the composites had occurred. As the impact energy increased, the peak force obtained also increased. All the specimens have a longer contact time at 90 J of impact energy than the lower impact energy levels. E400 specimens had a longer contact time than other specimens, which was up to almost 5 ms. A longer contact time indicates that the higher energy can be absorbed by the specimens and that the damaged area can become larger (Pantelakis et al. 2016). The E800 specimen at the impact energy of 90 J had the highest peak force among the other specimens. The lowest peak force was obtained by the E400 specimens whose peak force was only up to 8 kN at 15 J of impact energy.

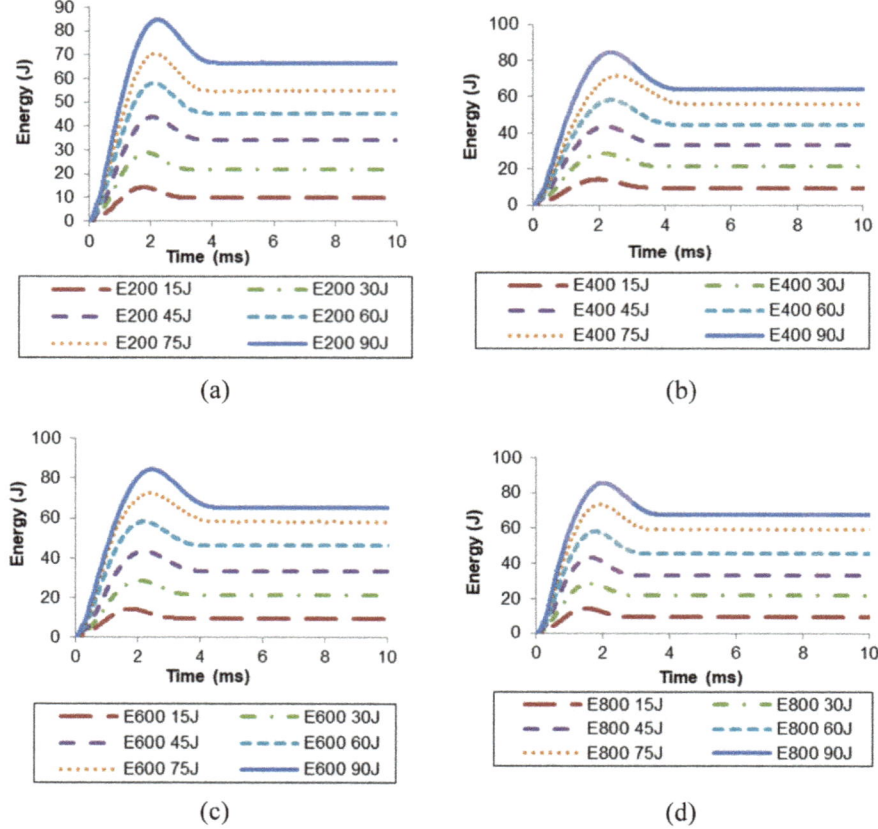

Fig. 2 Energy–time curves of the flax-kenaf fibre composite hybridised with glass fibre: **a** E200; **b** E400; **c** E600; and **d** E800

Figure 2 illustrates the energy against time data of the flax and kenaf fibre composite that were hybridised with the glass fibre. All the specimens showed a similar trend of energy-time curve. The energy gradually increased until the maximum impact energy was reached and began to drop until the energy reached a constant value. The time taken also increased as the impact energy was increased. The highest peak of the graph shows the impact energy. Energy absorb can be calculated from this graph as the absorbed energy is equal to energy at the constant value. Rebound energy can be determined as impact energy minus with absorbed energy. All graph shows that rebound occurs which means no penetration occurs during the impact tests. As the impact energy increase, the absorbed energy also increased. As the absorbed energy increase, damage occurs on the impacted specimens also has larger damage area.

To analyse the damage progression of the impact damage, the graph of force against displacement was studied. A closed curve from the graph indicates that the

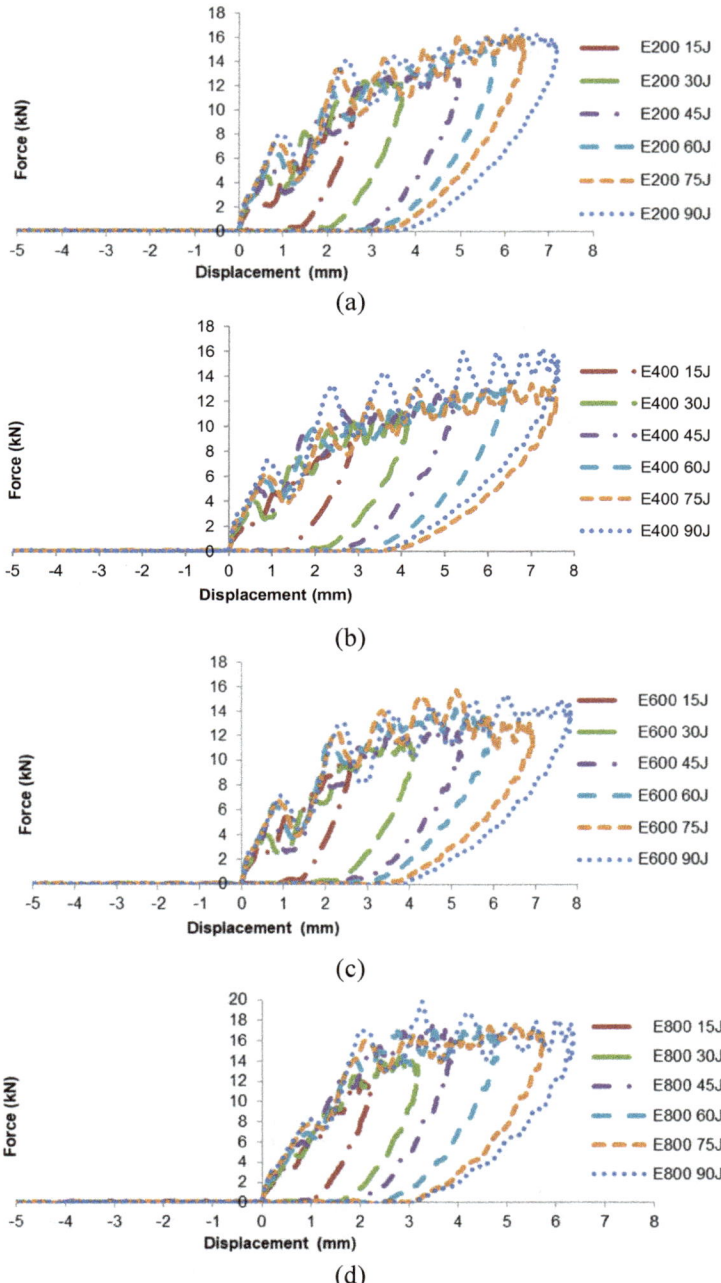

Fig. 3 Force–displacement curves of flax-kenaf fibre composite hybridised with glass fibre: **a** E200; **b** E400; **c** E600; **d** E800

striker did not penetrate the specimen during testing. However, an open curve from the same graph shows that the striker had penetrated the specimen. The maximum impact energy is shown as the highest tip of the curve while the end of the curve represents the absorbed energy (Ismail et al. 2018). Figure 3 clearly shows that all the specimens tested had closed curves. This indicates that none of the specimens was penetrated even at 90 J of impact energy. This means that this type of composites can withstand a higher impact and can further be investigated using high-velocity impact test. The contrast between the impact energy and the bounce back energy is the energy absorbed by the specimens. It compares to the zone encased between the stacking and emptying curves. The energy is absorbed because of the damage initiated in the specimens and the vibrations of the test machine and the striker. From the force–displacement curves, the energy absorbed also can be determined where it is equal to the area under the graph. Figure 4 illustrates the average absorbed energy from all the three tested specimens for all the types of composites.

As the impact energy increased, the energy absorbed by the specimens also increased. This shows a good correlation between the impact energy and its properties. It demonstrates that more damage occurs at a higher impact energy. From Fig. 4, it can be clearly seen that E600 had the highest absorbed energy at almost all impact energy levels. At 15 J of impact energy, E200 had the lowest energy absorbed, while the other three types of specimens had almost the similar value of energy absorbed. At 30 J of impact energy, the E600 specimens had absorbed slightly higher energy compared to the others. The E600 specimens maintained to have higher energy absorbed until 90 J of impact energy which reached 70 J of energy absorbed.

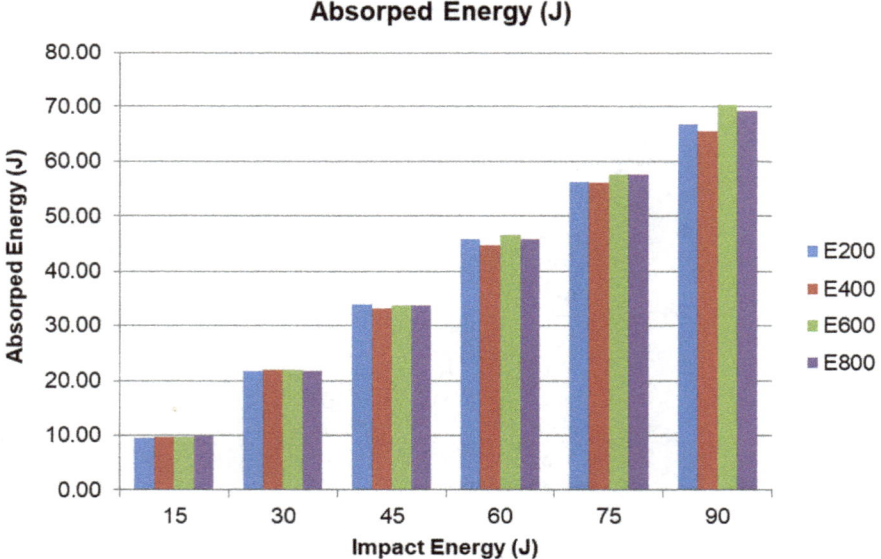

Fig. 4 Absorbed energy of the hybrid composites

E400 had the lowest energy absorbed at almost all levels of impact energy. This indicates that E400 has the smallest fracture compared to the others. From the data observation, the impact energy is higher than the energy absorbed. This indicates that the specimens were not penetrated even at the highest impact energy.

3.2 Compression After Impact

The compression after impact (CAI) was done to determine the strength of the composites after the low-velocity impact test. Figure 5 below illustrates the behaviour of the specimens of kenaf/flax/glass hybrid composites on CAI tests after the low-velocity impact test at different energy levels (15, 30, 45, 60, 75 and 90 J) with different types of glass fibre (E200, E400, E600 and E800).

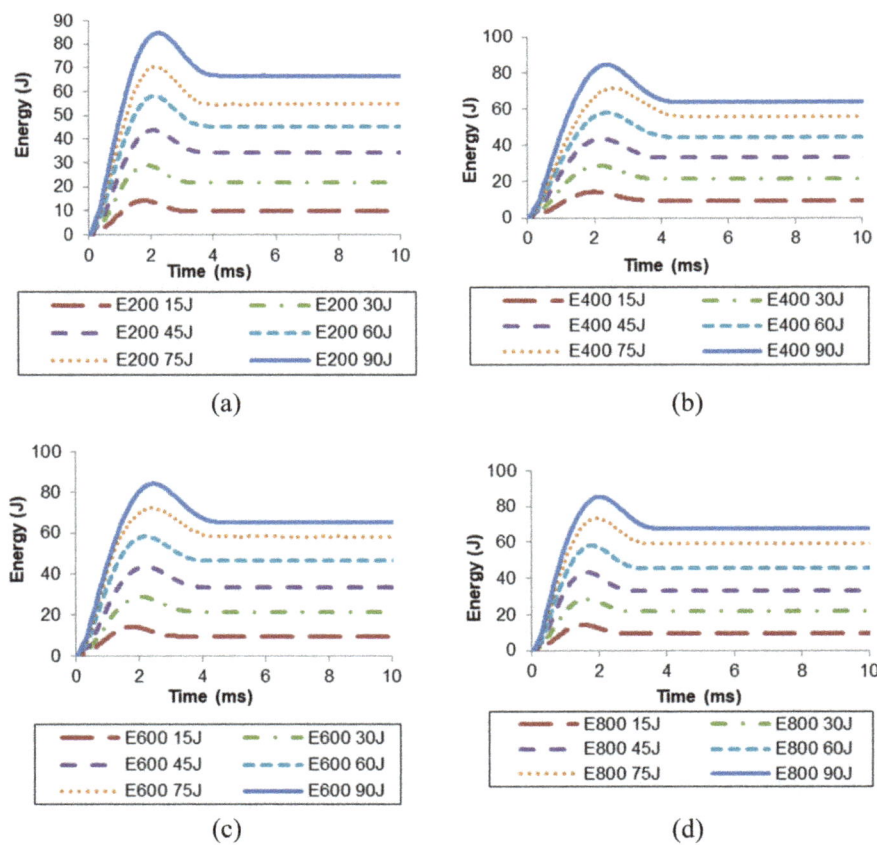

Fig. 5 Compressive stress–compressive strain curves: **a** E200; **b** E400; **c** E600; and **d** E800

From the results in Fig. 5, it can be observed that as the impact energy increased, the compressive stress decreased. The glass fibre type itself affects the compressive strength. The pattern of the curves is almost similar for all the specimens: as the compressive stress increased, the compressive strain also increased until it reached the maximum stress. After the maximum stress was achieved, the specimens began to fail and the value of the stress became constant and slowly dropped. The compression on the impacted specimens caused the specimens to fail and start to buckle on the rear surface. This led to a perpendicular delamination to the load direction. The fibre breakage and matrix cracking started to propagate at the area with higher stress concentration before the specimens failed. Fibre cracks were assumed to be detected just before the failure of specimens. From the graph, the highest compressive strength obtained was by the E200 specimens at 15 J of impact energy, which was almost 60 N/mm^2. This may due to the density and the crosslinking of the fibre itself that is higher than the other specimen types. However, the compressive strain obtained from the E200 specimens was much lower than the other specimens, which was less than 3% when it reached the maximum compressive strength. For the E400 and E600 specimens, the specimens began to fracture and the value of compressive stress began to drop after 3% of the compressive strain was obtained.

4 Conclusions

From the research, it tends to be observed that various sorts of material have various outcomes at low-velocity impact. The material properties influence the stiffness of the composites and the contact stiffness significantly affects the dynamic reaction of the composites. It is critical to comprehend the properties of damage after impact. This kind of impact creates a few methods of damage emerging inside, the material thickness with no observation on the impacted side, this damage are included because of the heterogeneity and the anisotropy of the composite. The damage occurs at the low-velocity impact on these composites were delamination, matrix splitting, and matrix breakage. Diverse impact energy levels brought about various errors and damage.

From the graph of force and displacement, the area below the graph represents the energy absorbed by the specimen. The impact energy that can be absorbed by the specimens shows that the specimens were capable to withstand the impact force. All force–displacement curves have closed loops which means the impactor did not penetrates the specimens as the occurrence energy was completely moved back to the samples when it reaches the maximum displacement. As none of the specimens had been penetrated, it shows that the composites can withstand much higher impact energy and impact velocity.

The results show that the E600 specimens had the highest energy absorption. Most of the E200 specimens only had matrix cracking failure and delamination. This statement is in accordance as the visual observation which showed that no specimens were penetrated and can further investigate by using another non-destructive test and destructive test. It can be concluded that the hybridization of these materials is safe to be implemented in structural applications.

Comparing the samples, it can be concluded that a larger damage area contributes to a lower compressive force in CAI testing. Visible damage could be seen upon inspection with the unaided eye, including cracks. It is well known that the specimens had previously undergone damage caused by the low velocity impact. Therefore, the compression after impact test increased the damage area and the cracks that had been initiated in the inner layers of the material, transferring to the front layers due to significant delamination. Hence, the reinforcement that increased the toughness and impact resistance of the composite matrix should be additionally examined.

Acknowledgements This work is supported by UPM under GP-IPS (Grant No. 9486400). The authors would like to thank the Department of Aerospace Engineering, Faculty of Engineering, Universiti Putra Malaysia and the Laboratory of Biocomposite Technology, Institute of Tropical Forestry and Forest Products (INTROP), Universiti Putra Malaysia (HICOE) for close collaboration in this research.

References

Agrawal S, Singh KK, Sarkar PK (2014) Impact damage on fibre-reinforced polymer matrix composite-a review. J Compos Mater 48. https://doi.org/10.1177/0021998312472217

Akin C, Senel M (2010) An experimental study of low velocity impact response for composite laminated plates. DPU Fen Bilimleri Enstitusu Dergisi, Sayi 21:77–90

Balali E, Kordani N, Vanini AS (2017) Response of glass fibre reinforced hybrid shear thickening fluid (STF) under low velocity impact. J Text Inst 108:376–384

Baley C, Gomina M, Breard J, Bourmaud A, Drapier S, Ferreira ML, Duigou A, Liotier PJ, Ouagne P, Soulat D, Davies P (2018) Specific features of flax fibres used to manufacture composite materials. Int J Mater Form. https://doi.org/10.1007/s12289-018-1455-y

Borvik T (2003) An introduction to impact and penetration dynamics. Department of Structural Engineering, Norwegian University of Science and Technology

Channabasavaraju S, Shivanand HK, Santhosh Kumar S (2013) Investigation of low velocity impact properties of kevlar fiber reinforced polymer matrix composites. Int J Eng Res Technol (IJERT) 02(10)

Cheng X, Zhao W, Liu S, Xu Y, Bao J (2014) Damage of scarf-repaired composite laminates subjected to low-velocity impacts. Steel Compos Struct 17:199–213. https://doi.org/10.12989/scs.2014.17.2.199

Duell JM (2014) Impact testing of advanced composites. Adv Topics Charact Compos 97–112

Ghelli D, Minak G (2011) Low velocity impact and compression after impact tests on thin carbon/epoxy laminates. Compos B Eng 42(7):2067–2079

Ismail MF, Sultan MT, Hamdan A, Shah AU (2018) A study on the low velocity impact response of hybrid kenaf-kevlar composite laminates through drop test rig technique. BioResources 3045–3060

Kaw AK (2006) Mechanics of composite materials, 2nd edn. CRC

Kim H, Halpin JC, DeFrancisci GK (2012) Impact damage of composite structures. In: Long-term durability of polymeric matrix composites, pp 143–180

Mohanty, AK, Misra M, Drzal LT (2005) Natural fibres, biopolymers, and biocomposites. CRC, Florida, USA. 2004058580

Pantelakis SG, Katsiropoulos CV, Polydoropoulou PV (2016) Assessing the compression after impact behaviour of innovative multifunctional composites. Nanomater Nanotechnol 6:184798041667962. https://doi.org/10.1177/1847980416679627

Pickering KLA, Efendy MG, Le TM (2016) A review of recent development in natural fibre composites and their mechanical performance. Compos A Appl Sci Manuf 83:98–112

Rafiq A, Merah N, Boukhili R, Al-Qadhi M (2017) Impact resistance of hybrid glass fibre reinforced epoxy/nanoclay composite. Polym Testing 57:1–11

Rahman MM, Hosur M, Hsiao KT, Wallace L, Jeelani S (2015) Low velocity impact properties of carbon nanofibres integrated carbon fibre/epoxy hybrid composites manufactured by OOA-VBO process. Compos Struct 120:32–40

Razali N, Sultan MTH, Jawaid M (2018) Impact damage analysis of hybrid composite materials. In: Durability and life pediction in biocomposites, fibre reinforced composites and hybrid composites. Woodhead Publishing, pp 121–132

Ricci F, Leece L, Monaco E, Maio L (2013) Simulation of velocity impact on composite laminates. Aerospace Engineering Department, University of Naples, Naples

Stronge WJ (2018) Role of impact in development of mechanics during the seventeenth and eighteenth centuries. In: Impact mechanics. Cambridge University Press, p 332

Tholibon D, Tharazi I, Sulong AB, Muhammad N, Ismail NF (2019) Kenaf fiber composites: a review on synthethic and biodegradable polymer matrix. Jurnal Kejuruteraan 31(1):65–76

Tirillo J, Ferrante L, Sarasini F, Lampani L, Barbero E, Sanchez-Saez S, Valente T, Gaudenzi P (2017) High velocity impact behaviour of hybrid basalt-carbon/epoxy composites. Compos Struct 168:305–312

Lightning Source UK Ltd.
Milton Keynes UK
UKHW020739270522
403612UK00001B/15

9 789811 613258